What If Change Has an Architecture?

Sandeep Chavan

Published by Gyrus Vision, 2026.

WHAT IF CHANGE HAS AN ARCHITECTURE?

First edition. August 26, 2026.

ISBN: 979-8235489493

Written by Sandeep Chavan.

Table of Contents

Part I | Familiar Events, Different Questions 1
Chapter 1 | Heat Without a Passenger 2
Chapter 2 | Difference Comes First........................ 18
Chapter 3 | The Boundary Matters........................ 33
Interlude I........................ 48
Part II | Beneath the Measured Quantities 51
Chapter 4 | Reading Thermal Condition 52
Chapter 5 | Pressure at the Edge........................ 67
Chapter 6 | Equilibrium Without Stillness........................ 82
Interlude II........................ 97
Part III | The U E D Layer, Carefully Entered 100
Chapter 7 | Persistence Within Continuity........................ 101
Chapter 8 | The Available Door 116
Chapter 9 | Entropy as Record 130
Interlude III 145
Part IV | Across Fields and Toward Research........................ 148
Chapter 10 | Machines Organize Pathways........................ 149
Chapter 11 | One Architecture, Many Fields........................ 164
Chapter 12 | From View to Science........................ 179
Closing Reflection........................ 194
Core Definitions........................ 198
A Suggested Reading Path 202
Selected Further Reading........................ 204
Acknowledgements........................ 206
About the Author........................ 208
Further Reading Within the Chavanian Ecosystem 210
Disclaimer and Statement of Scope........................ 212

For those who learned the equations, but never stopped asking what physical story they were telling.

"Difference begins the question.Architecture determines what the system can do with it."

— Sandeep Chavan

Opening Note

The Book Is an Invitation, Not a Replacement

Thermodynamics does not need rescuing. Its laws have survived engines, laboratories, chemical industries, weather models, living systems, spacecraft, and technologies that the earliest pioneers could not have imagined. When a refrigerator cools a compartment, when a turbine turns, when a metal changes phase, or when a living body regulates its temperature, thermodynamics continues to provide dependable guidance. That success is the starting point of this book, not an obstacle that the book intends to overcome.

The invitation here is simpler. What if we keep every established measurement and every successful relation in place, but allow ourselves to ask a different kind of question? What if we ask what sort of physical architecture permits a difference to become a change? What if temperature, pressure, heat, work, and entropy remain exactly as useful as science has shown them to be, while we explore a deeper picture of the conditions beneath their appearance?

That deeper picture is what I call the Chavanian view. It is not a replacement for thermodynamics. It is not a shortcut around statistical mechanics. It does not ask teachers to discard the language their students need, and it does not give researchers permission to ignore experimental definitions. It is a lens placed one conceptual layer beneath familiar descriptions. A lens may alter what we notice without altering the object being observed.

This distinction matters because thermodynamics often arrives in people's lives indirectly. A biologist meets it through metabolism and membranes. A materials scientist meets it through phase, stability, and processing. A chemist meets it through reaction, equilibrium, and chemical potential. An engineer meets it through machines, fluids, heat exchangers, and losses. An atmospheric scientist meets it through radiation, circulation, moisture, and gradients. Many of these readers do not need another long derivation. They need a coherent physical story that helps the established quantities sit together in the mind.

The story offered here begins with difference. A warmer region and a cooler region are not the same. A compressed gas and its surroundings are not in the same condition. Two regions with different concentrations are not interchangeable. In the language used throughout this book, such a relevant difference is called Delta. Delta is not a substance, a force, or a hidden supply of energy. It is simply a way of drawing attention to a relation that makes some form of change possible.

A difference alone does not specify what will happen. A wall may conduct or insulate. A piston may move or remain locked. A membrane may permit one species to pass while stopping another. A crystal may preserve its structure until a threshold is crossed. A living system may spend resources to maintain a difference that would otherwise diminish. The available pathway matters as much as the initiating condition.

This is where the word architecture enters. Architecture does not mean a building hidden inside matter. It means the organized set of boundaries, capacities, connections, constraints, and possible routes through which a system can respond. The same broad intervention can produce different outcomes because the architecture is different. Heat a gas beneath a movable piston and the response can include expansion. Heat the same kind of gas in a rigid container and the response must be accommodated differently.

Teachers may recognize that this is often what students are trying to understand when they ask apparently simple questions. Where did the heat go? Why did the temperature stop rising? Why does a refrigerator need work? Why can two objects have the same temperature but contain very different amounts of internal energy? The equations may answer the quantitative problem while the student is still searching for a physical narrative.

Early researchers encounter a related difficulty. They learn to calculate within a domain, yet the words used across disciplines do not always line up neatly. Energy may be described as stored, released, dissipated, available, unavailable, embodied, transported, or consumed. Each expression can be operationally useful, but the images they create may conflict. A conversational framework can help separate the dependable calculation from the metaphor used to imagine it.

The Chavanian view proposes that difference, persistence, boundary, accessibility, resolution, consequence, and changed condition belong to one recurring story. The story is deliberately modest in this book. It will not be presented as a completed physical theory. Universal Energy Dynamics, or U E D, will appear as a proposed foundational setting in which continuity is taken seriously. It will remain clearly marked as proposal rather than consensus.

You will find no equations in the pages that follow. That is not because equations are unimportant. It is because this book has another job. It is meant to be read aloud, listened to, discussed in a classroom, carried into a laboratory conversation, or opened by a scientist whose main field lies elsewhere. It is a companion to technical study, not a substitute for it.

The most important rule is therefore simple. Whenever established thermodynamics gives a tested description, that description remains. Whenever the Chavanian view offers an interpretation, it will be named as a view. Whenever U E D reaches beyond what has been experimentally established, the boundary will be stated. Curiosity does not become stronger by hiding uncertainty. It becomes stronger by knowing exactly where the question begins.

So this book starts with familiar things: a cup cooling on a desk, a wall holding back a gas, a refrigerator humming in a kitchen, a material changing phase, a cell maintaining a gradient. We will look at them carefully and ask a recurring question. What if change has an architecture? The answer will not replace thermodynamics. It may, however, help us see why thermodynamics belongs to so many different fields at once.

Preface

Why Write a Conversational Companion?

The larger work called Chavanian Thermodynamics was written to examine foundational questions in detail. It moved slowly through systems, persistence, boundaries, temperature, pressure, heat, work, entropy, equilibrium, statistical mechanics, machines, low-temperature limits, and testability. That scale was necessary for the argument it was trying to make. It also made the book demanding. A reader could agree that the laws were not under attack and still feel that the conceptual route required more time than an ordinary teaching term, research schedule, or interdisciplinary curiosity could offer.

This companion has a different purpose. It does not compress the larger work chapter by chapter, and it does not attempt to preserve every distinction developed there. It asks what a teacher, student, early researcher, or working scientist most needs in order to enter the Chavanian view without being buried by it. The answer is not a catalogue of terms. It is a sequence of familiar questions.

What if heat is not imagined as a travelling material? What if a difference is the beginning of change but not its cause in the ordinary mechanical sense? What if a boundary participates in the outcome? What if equilibrium is not featureless sameness? What if entropy can be discussed without turning disorder into an invisible agent? What if engines and living systems both manage pathways without becoming the same kind of system?

These questions are useful because they begin in experience. A teacher can ask them before introducing a formal definition. A student can return to them after completing a calculation. A researcher can use them to notice where disciplinary language has quietly become metaphysical. A scientist from another field can follow the physical narrative without first mastering a new notation.

The conversational tone is deliberate. Foundational questions are often made difficult twice: first by the difficulty of the subject and then by the pressure to sound technical before the idea is ready. Technical language is indispensable when it carries a precise distinction. It is less helpful when it protects vagueness. In this book, an ordinary sentence must do the work before a specialized term is allowed to enter.

That does not mean the science will be treated casually. The Zeroth, First, Second, and Third Laws remain part of the background. So do kinetic theory, statistical mechanics, phase behaviour, chemical thermodynamics, and non-equilibrium processes. They will be described in words rather than developed through mathematical derivation. Readers who need the quantitative structure should use standard textbooks and primary literature alongside this book.

The phrase Chavanian view also requires care. It identifies a proposed way of organizing familiar observations around continuity, persistence, Delta, boundaries, available pathways, resolution, consequence, and changed geometry. A view can be coherent and useful without being experimentally distinct. If it eventually produces new quantitative predictions, it can be tested more strongly. Until then, its main contribution is conceptual organization.

Universal Energy Dynamics sits beneath this view as a provisional physical proposal. It suggests that the world should not begin conceptually with isolated objects placed in an unrelated emptiness. It begins instead with a continuous foundational reality within which matter appears as persistent structured manifestation. This book will explain why that picture attracts the Chavanian framework. It will also state plainly that Structured Vacuum Energy and Alignment Threshold Time are not established components of thermodynamics.

The book therefore moves in stages. First we look again at everyday thermal events. Then we consider what familiar quantities measure and what they do not need to become in our imagination. After that we enter the U E D layer carefully. Finally, we carry the view across machines, chemistry, materials, biology, atmosphere, and research practice.

Each chapter returns to a reading boundary. What does established science say? What does the Chavanian view add? What remains provisional? The repetition is intentional. A speculative framework becomes safer when the reader never has to guess which level a statement belongs to.

This structure also makes the manuscript suitable for narration. Important terms are introduced in full. Symbolic expressions are avoided. Visual diagrams are not required. Each section develops one main movement and ends by returning to the chapter question. The listener should be able to pause, resume, and still know where the discussion stands.

The desired result is neither a popular slogan nor a hidden textbook. It is a bridge. On one side lies the dependable practice of thermodynamics. On the other lies a proposed architecture beneath that practice. The bridge allows conversation between them while preventing one side from being mistaken for the other.

If the book succeeds, a student will finish it with more respect for thermodynamics, not less. A teacher will gain another way to open difficult discussions. A researcher will see new questions without confusing them with results. And a scientist from another field will understand why the same thermodynamic themes keep appearing wherever systems persist, encounter differences, and change.

Introduction

How to Read a View Without Mistaking It for a Replacement

Imagine three transparent layers placed over the same physical event. On the first layer is observation. A cup cools. A gas expands. A membrane maintains a concentration difference. A crystal changes phase. These are events that can be measured, repeated, and described with increasing precision.

On the second layer is established thermodynamics. Here we define systems and surroundings, identify boundaries, distinguish state quantities from process quantities, and connect measurements through tested relations. This layer is extraordinarily successful. It gives engineering its discipline and allows scientists in different laboratories to compare results.

On the third layer is interpretation. This is where we ask what kind of physical story we attach to the measurements. Is heat pictured as something that travels? Is energy imagined as a content residing inside an object? Is entropy treated as disorder, missing information, state multiplicity, dispersal, or a thermodynamic coordinate? Different explanatory traditions can share the same mathematics while emphasizing different physical pictures.

The Chavanian view belongs primarily to this third layer. It proposes that many thermodynamic events can be narrated through difference, available pathway, resolution, consequence, and changed condition. It asks us to see the boundary as part of the event and to treat persistence as something that must be explained rather than assumed.

Beneath that interpretive layer lies a further proposal: Universal Energy Dynamics. U E D begins from continuity and introduces Structured Vacuum Energy as a foundational substrate. Matter, in this view, is persistent structure within that continuity. Alignment Threshold Time is proposed as part of the account of when a pathway becomes physically accessible. These ideas do not presently belong to established thermodynamics, and this book will not present them as though they do.

A useful reading habit is to ask which layer a sentence occupies. If the sentence says that temperature is measured, that pressure is exerted at a boundary, or that a spontaneous process has a thermodynamic direction under stated conditions, it belongs to established science. If the sentence says that these events may be viewed as the resolution of Delta through an available geometry, it belongs to the Chavanian interpretation. If the sentence locates the event in Structured Vacuum Energy, it belongs to the proposed U E D foundation.

The word Delta is used because ordinary words such as difference, gradient, imbalance, and driving force carry different technical histories. Delta in this book is intentionally broad. It means a relational difference that is relevant to what a system can do. Not every difference is immediately active. A wall may block a route. A threshold may not be reached. A competing process may dominate. Delta names the condition; it does not act like a little agent pushing the system.

Resolution means the physical process through which an available difference ceases, diminishes, changes form, becomes redistributed, or creates a new condition. The word must not be confused with disappearance. A difference can resolve locally while producing another difference elsewhere. A refrigerator maintains a cool interior by creating consequences in a wider system. A living organism preserves internal differences by drawing upon and altering its surroundings.

Architecture means the pattern of possible and blocked routes. It includes boundaries, material properties, degrees of freedom, phase, scale, coupling, and history. In a classroom example, architecture may be as simple as whether a piston can move. In a cell, it may involve membranes, enzymes, transport proteins, chemical networks, and feedback. In an atmosphere, it may include radiation, convection, rotation, moisture, and geography.

Persistence refers to a structure or condition that remains identifiable through time despite ongoing internal activity. A molecule persists. A vortex may persist. A living cell persists only through continuous exchange and regulation. A thermodynamic state can be treated as stable over the timescale relevant to a measurement even while microscopic events continue.

These terms are not offered as replacements for technical vocabulary. They are organizing terms. A chemist should still speak precisely about enthalpy, chemical potential, activity, phase, and equilibrium. An engineer should still distinguish heat transfer from work and apply the correct sign convention. A biologist should still use the established language of metabolism, transport, and regulation. The Chavanian terms sit around those practices as a conceptual frame.

The book is also careful about scale. A description that is illuminating for a gas in a cylinder may become misleading in a turbulent atmosphere or a living cell. Similar architecture does not mean identical mechanism. The fact that two systems both manage gradients does not make them the same kind of system. The comparison is useful only when the differences remain visible.

For teachers, each chapter can be treated as a discussion prompt. Begin with the opening question. Invite students to describe the event in ordinary language. Then identify which statements are measurements, which are established relations, and which are interpretations. This exercise often reveals that confusion enters not through the calculation but through an unnoticed shift between layers.

For students, the book is best read alongside formal study. When a technical course introduces a state function, process path, phase transition, or entropy balance, return to the corresponding chapter and ask what picture helps the relation make sense. If the picture conflicts with the formal definition, keep the formal definition and revise the picture.

For researchers, the final chapters are the most important. A conceptual framework earns scientific weight by becoming constrained. It must define terms operationally, recover known results, identify where it differs, and accept the risk of being wrong. The Chavanian view is presented here as an organized research direction, not as a completed rival theory.

For listeners, no visual memory is assumed. When a term matters, it will be restated in words. When a chapter crosses from established science to interpretation, the transition will be spoken plainly. The central questions are designed to remain meaningful without a page in front of you.

We can now begin with the most familiar thermodynamic word of all: heat. We will not deny the measurement, discard the process, or argue with the thermometer. We will ask what kind of story we tell when one condition changes in the presence of another, and whether the story becomes clearer when change is understood as structured resolution rather than the journey of an invisible substance.

Part I

Familiar Events, Different Questions

Chapter 1
Heat Without a Passenger

What If Heat Is Not a Travelling Substance?

The Cup Cools, but What Is the Story?

A cup of tea is placed on a desk. At first it is too hot to drink. Later it is comfortable, and later still it approaches the temperature of the room. Everyone recognizes the event. The difficulty begins only when we try to say what moved, what changed, and what the word heat is asking us to imagine.

The question in this chapter is not whether established thermodynamics works. It is what we notice when we place the idea of architecture beside heat, temperature change, and thermal contact. The calculation and the measurement remain where science has placed them. Our task is to build a physical narrative that is clear enough for teaching, useful enough for interdisciplinary conversation, and modest enough to preserve the difference between interpretation and evidence.

We will move through familiar scenes, the established account, the Chavanian view, and the limits of that view. By the end, the phrase heat names a process, not a resident material should feel less like a slogan and more like a disciplined question that can accompany technical study without competing with it.

The Cooling Cup

Consider what an observer would actually notice. A warm cup, the cooler air, the table, and the nearby walls form one thermal situation. The cup changes while the room changes so slightly that we usually ignore the room's response. None of this observation tells us, by itself, which metaphor must be used underneath it. The event gives us readings, timings, boundaries, and consequences. The deeper story is something we add in order to make those relations intelligible.

Thermodynamics and the relevant specialist science already organize the event. Established thermodynamics treats heat as energy transferred across a boundary because of a temperature difference. The cup's internal energy changes, and the details can involve conduction, convection, radiation, evaporation, and the thermal properties of the materials. These established descriptions are the common ground. They allow a teacher in one country, an engineer in another, and a researcher in a third field to speak about the same physical situation without depending on personal imagery.

Now introduce the Chavanian lens carefully. The Chavanian view describes a thermal Delta between the cup and its surroundings and asks how the coupled geometry allows that difference to resolve. The process is real without requiring heat to be imagined as a fluid stored in the tea. Instead of asking where an invisible content went, the lens asks how the system's condition, boundary, and available degrees of freedom shaped the response. The change of language is useful only if it produces greater discipline rather than a new ambiguity.

This way of speaking travels across disciplines with caution. The same discipline helps when discussing a warm animal in cold air, a hot component on a circuit board, or a lake exchanging thermal consequence with the atmosphere. The recurring architecture may be similar even when the mechanisms, scales, and measurable variables are very different. A membrane is not a piston, a cell is not an engine, and an atmosphere is not a sealed laboratory vessel. Analogy helps only when it does not erase those distinctions.

A teacher can turn this into a conversation. Ask learners to separate the temperature readings from the process word heat. Then ask them to identify every boundary through which the event can occur and every pathway that would change if the cup were covered. Ask which statement came from observation, which came from a tested model, and which came from interpretation. Students often discover that they understood the formal relation but had mixed several physical pictures together. Naming the layers gives them permission to separate those pictures without discarding the science.

The research boundary is equally important. The phrase thermal resolution does not replace heat transfer coefficients, energy balances, or microscopic explanations. It offers a narrative around them. U E D does not gain experimental status merely because it offers a coherent vocabulary. Structured Vacuum Energy and Alignment Threshold Time require independent definitions, connections to existing measurements, and possible tests. The present book keeps that obligation visible.

What, then, should remain after this section? The cup does not lose a material called heat. It participates in a thermal process through which an initially relevant difference becomes smaller and the combined surroundings enter a slightly changed condition. The Chavanian view invites another question after the standard explanation is complete: what architecture made this particular thermodynamic response possible?

Let us return to the event without technical haste. A warm cup, the cooler air, the table, and the nearby walls form one thermal situation. The cup changes while the room changes so slightly that we usually ignore the room's response. An observer could record what happens, but a fuller explanation would also ask what had to be present before it could happen. There had to be a particular arrangement, a particular contact with the surroundings, and a particular range of possible responses. The event is therefore not only a change in a quantity. It is a change made possible by a physical situation.

This is where ordinary grammar can quietly mislead us. The cup does not lose a material called heat. It participates in a thermal process through which an initially relevant difference becomes smaller and the combined surroundings enter a slightly changed condition. A convenient sentence may give an abstract quantity the role of an actor, as though the quantity itself travelled, commanded, or chose. Scientists usually understand the shorthand, but new learners may build their intuition around the grammar. Rephrasing the sentence in terms of conditions, interactions, and measurable change often reveals what the original expression had compressed.

Sunlight, Skin, and a Metal Bench

A useful way to enter the topic is to slow the event down in imagination. A person steps from shade into sunlight and feels warmer. The same person touches a metal bench and may experience it as colder than nearby wood even when both have been in the same environment. At each moment, the system has a condition and a set of possible next steps. Some steps are easy, some require a threshold, and some are excluded by the way the system is built or connected.

The established account tells us how to describe that condition responsibly. Radiative absorption, thermal conductivity, heat capacity, contact area, blood flow, and sensory response all shape the experience. Temperature and the rate of thermal exchange are related but not identical. It protects us from replacing measurement with intuition. Even in an equation-free conversation, the formal background matters because the words inherit their meaning from that disciplined practice.

From the Chavanian perspective, the event can be narrated as structured resolution. The Chavanian view asks how each material and boundary architecture makes a different route available for the same body. The sensation is not a direct reading of a substance called cold or heat. Delta names the relevant difference. The architecture names the permitted and blocked routes. Resolution names the actual process. Consequence names the changed condition that remains after the process has occurred.

That sequence can illuminate another field. Materials selection in buildings, protective clothing, cookware, medical devices, and spacecraft depends on managing these different routes rather than merely choosing a desired temperature label. Yet the illumination is conceptual rather than a claim that one universal mechanism has already been demonstrated. The value lies in asking comparable questions: What persists? What differs? What pathway is available? What must the surroundings supply or receive?

In a classroom or research group, the vocabulary can be used as a checking device. A useful discussion begins by asking why equal thermometer readings need not produce equal sensations. The answer opens the distinction between state, process, capacity, and pathway. If the conversation begins to treat Delta as a force, architecture as a hidden machine, or resolution as a new measurable substance, the terms have drifted. Their purpose is to organize relations, not to populate nature with additional invisible objects.

The strongest caution is this. Human thermal sensation includes physiology and perception. It should not be reduced to a single thermodynamic variable or used as a substitute for measurement. Established thermodynamics can stand without the Chavanian interpretation. The interpretation must therefore earn its place through explanatory usefulness now and through mathematical and experimental work later. It cannot demand immunity from the evidence that shaped the field.

The chapter's central movement appears again. What a body experiences depends on the relation between two systems and the pathway joining them, not on temperature alone. Change becomes easier to discuss when we keep both the initiating difference and the available pathway in view. The answer is not hidden in either one alone.

Imagine watching the scene with the ability to stop it at several moments. A person steps from shade into sunlight and feels warmer. The same person touches a metal bench and may experience it as colder than nearby wood even when both have been in the same environment. At every pause, the system presents both a current condition and a limited range of next conditions. The range is shaped by material structure, contact, geometry, and surrounding constraints. Change is therefore neither a detached command nor a mysterious jump. It occurs through what the present arrangement can physically support.

A small change in wording can expose this structure. What a body experiences depends on the relation between two systems and the pathway joining them, not on temperature alone. Instead of asking what substance moved, ask which relation changed. Instead of asking what quantity wanted to increase or decrease, ask which processes were permitted under the stated conditions. The older wording may still be retained for convenience, but the second wording makes its hidden assumptions easier to inspect. That is especially helpful when familiar metaphors begin to conflict with one another.

When Heating Does Not Raise Temperature

Begin with the scene itself. Ice can receive thermal input while melting, and boiling liquid can continue to receive thermal input while its temperature changes little under controlled conditions. The event surprises anyone who equates heating directly with becoming hotter. It is easy to hurry past such an event because it is familiar. Yet familiarity often hides the number of conditions that must cooperate. The materials have properties. The surroundings have a history. A boundary permits some exchanges and suppresses others. The event looks simple only after all of those details have already been arranged.

Established science gives us the disciplined account. During a phase transition, thermal input is associated with structural reorganization and latent heat. The temperature behaviour depends on pressure, composition, purity, and the path through the phase region. This account does not need a new vocabulary in order to remain valid. It tells us what to measure, which distinctions matter, and how the event can be compared with another event under specified conditions. Any deeper view must begin by respecting that achievement.

The Chavanian question begins one layer lower. The Chavanian view says that the system has acquired another accessible mode of accommodation. The thermal process now appears substantially as collective reconfiguration rather than only as a rising temperature coordinate. The word architecture directs attention away from an imagined travelling substance and toward the relations that make one response available while another remains blocked. This is an interpretation of the event, not a change to the event's measured description.

The same pattern becomes visible elsewhere. Phase-change materials used for thermal storage, cloud formation, metallurgy, food processing, and cryogenic systems all depend on this separation between thermal process and simple temperature rise. The comparison does not claim that the systems share an identical mechanism. It asks whether difference, constraint, pathway, and consequence provide a useful family resemblance across fields that often speak in separate dialects.

For teaching and early research, this distinction can be practical. Ask what changed when the thermometer barely moved. Students can name structure, fraction of phases, volume, density, and molecular organization before returning to the formal treatment. A learner can state the observation first, the accepted scientific relation second, and the proposed interpretation third. That order reduces the risk that a helpful picture will be remembered as though it were the definition itself.

There is also a limit that should remain audible. Latent heat is an established measurable quantity. The language of accommodation is interpretive and does not alter its definition. A conceptual account becomes scientifically stronger only when its terms can be constrained by measurement and when it accepts the possibility of failure. Until then, clarity is valuable, but clarity is not confirmation.

The section therefore returns to the chapter question. Thermal input does not carry a predetermined visible outcome. The material's accessible architecture helps decide whether the consequence appears as temperature change, phase change, expansion, or several responses together. In the Chavanian view, a change is not understood only by naming what changed. We also ask which difference mattered, which route was open, and what new condition the response created.

Pause once more over the ordinary details. Ice can receive thermal input while melting, and boiling liquid can continue to receive thermal input while its temperature changes little under controlled conditions. The event surprises anyone who equates heating directly with becoming hotter. Now imagine changing only one condition at a time: the material, the size, the surrounding medium, the duration of observation, or the character of the boundary. The visible event might remain recognizable, yet its pace and outcome could alter greatly. This is why a thermodynamic story becomes more accurate when it includes the setting instead of presenting the system as though it acted alone.

Language deserves its own small audit here. Thermal input does not carry a predetermined visible outcome. The material's accessible architecture helps decide whether the consequence appears as temperature change, phase change, expansion, or several responses together. In everyday speech we often turn a process into a thing. We say that heat enters, pressure pushes, or equilibrium arrives. Such expressions are efficient, and there is no need to ban them. The useful habit is to ask what the noun stands for in the actual account: a measured relation, an interaction, a redistribution, a constraint, or a shorthand for several of these together.

Heat and Temperature Are Not Synonyms

Consider what an observer would actually notice. A bathtub of warm water and a small cup of hotter water can reverse our casual expectations about which system is thermally more substantial. One has the higher temperature; the other may participate in a much larger thermal process. None of this observation tells us, by itself, which metaphor must be used underneath it. The event gives us readings, timings, boundaries, and consequences. The deeper story is something we add in order to make those relations intelligible.

Thermodynamics and the relevant specialist science already organize the event. Temperature is a state coordinate connected to thermal equilibrium and microscopic distributions. Heat is a process quantity associated with transfer across a chosen boundary because of temperature difference. These established descriptions are the common ground. They allow a teacher in one country, an engineer in another, and a researcher in a third field to speak about the same physical situation without depending on personal imagery.

Now introduce the Chavanian lens carefully. The Chavanian view keeps the distinction and adds a narrative: temperature helps describe the thermal condition, while heat names part of the measurable consequence occurring as a thermal Delta resolves across a boundary. Instead of asking where an invisible content went, the lens asks how the system's condition, boundary, and available degrees of freedom shaped the response. The change of language is useful only if it produces greater discipline rather than a new ambiguity.

This way of speaking travels across disciplines with caution. The distinction matters in medicine, climate science, industrial safety, cooking, battery management, and every field where a measured temperature can be mistaken for the total thermal significance of a system. The recurring architecture may be similar even when the mechanisms, scales, and measurable variables are very different. A membrane is not a piston, a cell is not an engine, and an atmosphere is not a sealed laboratory vessel. Analogy helps only when it does not erase those distinctions.

A teacher can turn this into a conversation. Have students explain why a high temperature does not tell them the mass, composition, phase, or total internal condition of a sample. The missing information reveals why state descriptions require several coordinates. Ask which statement came from observation, which came from a tested model, and which came from interpretation. Students often discover that they understood the formal relation but had mixed several physical pictures together. Naming the layers gives them permission to separate those pictures without discarding the science.

The research boundary is equally important. No single ordinary-language metaphor captures all uses of thermal quantities. Technical definitions must govern whenever precision is required. U E D does not gain experimental status merely because it offers a coherent vocabulary. Structured Vacuum Energy and Alignment Threshold Time require independent definitions, connections to existing measurements, and possible tests. The present book keeps that obligation visible.

What, then, should remain after this section? Temperature describes a condition. Heat describes a boundary process. Confusing them turns a useful calculation into an unreliable physical picture. The Chavanian view invites another question after the standard explanation is complete: what architecture made this particular thermodynamic response possible?

Let us return to the event without technical haste. A bathtub of warm water and a small cup of hotter water can reverse our casual expectations about which system is thermally more substantial. One has the higher temperature; the other may participate in a much larger thermal process. An observer could record what happens, but a fuller explanation would also ask what had to be present before it could happen. There had to be a particular arrangement, a particular contact with the surroundings, and a particular range of possible responses. The event is therefore not only a change in a quantity. It is a change made possible by a physical situation.

This is where ordinary grammar can quietly mislead us. Temperature describes a condition. Heat describes a boundary process. Confusing them turns a useful calculation into an unreliable physical picture. A convenient sentence may give an abstract quantity the role of an actor, as though the quantity itself travelled, commanded, or chose. Scientists usually understand the shorthand, but new learners may build their intuition around the grammar. Rephrasing the sentence in terms of conditions, interactions, and measurable change often reveals what the original expression had compressed.

A Process Without a Passenger

A useful way to enter the topic is to slow the event down in imagination. In daily speech we say that heat enters, leaves, escapes, or is trapped. These verbs are convenient because they make a process easy to picture, yet they also invite us to imagine a passenger moving between objects. At each moment, the system has a condition and a set of possible next steps. Some steps are easy, some require a threshold, and some are excluded by the way the system is built or connected.

The established account tells us how to describe that condition responsibly. Thermodynamics can quantify heat transfer without identifying heat as a material substance. Microscopic accounts describe interactions, motion, radiation, fields, collisions, and collective modes according to the system. It protects us from replacing measurement with intuition. Even in an equation-free conversation, the formal background matters because the words inherit their meaning from that disciplined practice.

From the Chavanian perspective, the event can be narrated as structured resolution. The Chavanian view treats the process as redistribution and changed condition within a connected physical situation. It asks us to track what systems do rather than search for a hidden thermal object. Delta names the relevant difference. The architecture names the permitted and blocked routes. Resolution names the actual process. Consequence names the changed condition that remains after the process has occurred.

That sequence can illuminate another field. Similar language problems occur when people say that a battery stores electricity, that pressure wants to escape, or that entropy causes disorder. Useful shorthand can quietly become an imagined ontology. Yet the illumination is conceptual rather than a claim that one universal mechanism has already been demonstrated. The value lies in asking comparable questions: What persists? What differs? What pathway is available? What must the surroundings supply or receive?

In a classroom or research group, the vocabulary can be used as a checking device. Invite learners to retell a thermal event without using the verbs contains, carries, or consumes. The exercise reveals which parts of their explanation are measurements and which parts are metaphor. If the conversation begins to treat Delta as a force, architecture as a hidden machine, or resolution as a new measurable substance, the terms have drifted. Their purpose is to organize relations, not to populate nature with additional invisible objects.

The strongest caution is this. Everyday language need not be banned. The goal is awareness, not verbal policing. A metaphor is safe when the user remembers that it is a metaphor.

Established thermodynamics can stand without the Chavanian interpretation. The interpretation must therefore earn its place through explanatory usefulness now and through mathematical and experimental work later. It cannot demand immunity from the evidence that shaped the field.

The chapter's central movement appears again. A process can cross a boundary without becoming a passenger. The event is the coordinated change of systems that were thermally different and physically connected. Change becomes easier to discuss when we keep both the initiating difference and the available pathway in view. The answer is not hidden in either one alone.

Imagine watching the scene with the ability to stop it at several moments. In daily speech we say that heat enters, leaves, escapes, or is trapped. These verbs are convenient because they make a process easy to picture, yet they also invite us to imagine a passenger moving between objects. At every pause, the system presents both a current condition and a limited range of next conditions. The range is shaped by material structure, contact, geometry, and surrounding constraints. Change is therefore neither a detached command nor a mysterious jump. It occurs through what the present arrangement can physically support.

A small change in wording can expose this structure. A process can cross a boundary without becoming a passenger. The event is the coordinated change of systems that were thermally different and physically connected. Instead of asking what substance moved, ask which relation changed. Instead of asking what quantity wanted to increase or decrease, ask which processes were permitted under the stated conditions. The older wording may still be retained for convenience, but the second wording makes its hidden assumptions easier to inspect. That is especially helpful when familiar metaphors begin to conflict with one another.

Closing Reflection

The question was not whether heat is real. Thermal processes are measurable, consequential, and indispensable to science. The question was whether the word must be imagined as a substance that resides in one body and then travels to another. Established thermodynamics does not require that picture. The Chavanian view prefers to speak of thermal Delta, available pathways, resolution, and changed condition.

Heat remains the established measure of a thermal boundary process. The Chavanian view asks us to understand that process as structured change rather than the journey of an invisible material.

Chapter 2

Difference Comes First

What If Difference Is the Beginning of Change?

Before Anything Happens, Something Is Not the Same

Before the cup cools, the cup and the room are thermally different. Before gas expands against a piston, a mechanical relation exists across the boundary. Before a dissolved substance diffuses, its distribution is not uniform. Change begins in a world that is already differentiated.

The question in this chapter is not whether established thermodynamics works. It is what we notice when we place the idea of architecture beside temperature, pressure, concentration, and other relevant differences. The calculation and the measurement remain where science has placed them. Our task is to build a physical narrative that is clear enough for teaching, useful enough for interdisciplinary conversation, and modest enough to preserve the difference between interpretation and evidence.

We will move through familiar scenes, the established account, the Chavanian view, and the limits of that view. By the end, the phrase Delta names a condition of possible change without acting as an agent should feel less like a slogan and more like a disciplined question that can accompany technical study without competing with it.

A Difference That Matters

A useful way to enter the topic is to slow the event down in imagination. Two objects can differ in colour without any thermal event following, while a small temperature difference can become relevant as soon as thermal contact is established. Not every difference belongs to every process. At each moment, the system has a condition and a set of possible next steps. Some steps are easy, some require a threshold, and some are excluded by the way the system is built or connected.

The established account tells us how to describe that condition responsibly. Physics identifies specific gradients, potentials, state differences, and constraints according to the domain. Their operational meanings are not interchangeable. It protects us from replacing measurement with intuition. Even in an equation-free conversation, the formal background matters because the words inherit their meaning from that disciplined practice.

From the Chavanian perspective, the event can be narrated as structured resolution. Delta is a broad Chavanian label for a relational difference that changes what resolutions are possible in a defined system. It does not erase the technical identity of temperature, pressure, concentration, voltage, or chemical potential. Delta names the relevant difference. The architecture names the permitted and blocked routes. Resolution names the actual process. Consequence names the changed condition that remains after the process has occurred.

That sequence can illuminate another field. A clinician, electrochemist, geologist, and atmospheric scientist may all work with gradients, yet each gradient belongs to a different measured structure and mechanism. Yet the illumination is conceptual rather than a claim that one universal mechanism has already been demonstrated. The value lies in asking comparable questions: What persists? What differs? What pathway is available? What must the surroundings supply or receive?

In a classroom or research group, the vocabulary can be used as a checking device. Ask what comparison is being made, which variable differs, and why the difference becomes relevant only under particular contact or coupling conditions. If the conversation begins to treat Delta as a force, architecture as a hidden machine, or resolution as a new measurable substance, the terms have drifted. Their purpose is to organize relations, not to populate nature with additional invisible objects.

The strongest caution is this. Delta must not become a universal number or vague cause. Without a defined system, variable, scale, and pathway, the word explains nothing. Established thermodynamics can stand without the Chavanian interpretation. The interpretation must therefore earn its place through explanatory usefulness now and through mathematical and experimental work later. It cannot demand immunity from the evidence that shaped the field.

The chapter's central movement appears again. Difference begins the question, but the kind of difference and the architecture that encounters it determine whether the question becomes a physical event. Change becomes easier to discuss when we keep both the initiating difference and the available pathway in view. The answer is not hidden in either one alone.

Imagine watching the scene with the ability to stop it at several moments. Two objects can differ in colour without any thermal event following, while a small temperature difference can become relevant as soon as thermal contact is established. Not every difference belongs to every process. At every pause, the system presents both a current condition and a limited range of next conditions. The range is shaped by material structure, contact, geometry, and surrounding constraints. Change is therefore neither a detached command nor a mysterious jump. It occurs through what the present arrangement can physically support.

A small change in wording can expose this structure. Difference begins the question, but the kind of difference and the architecture that encounters it determine whether the question becomes a physical event. Instead of asking what substance moved, ask which relation changed. Instead of asking what quantity wanted to increase or decrease, ask which processes were permitted under the stated conditions. The older wording may still be retained for convenience, but the second wording makes its hidden assumptions easier to inspect. That is especially helpful when familiar metaphors begin to conflict with one another.

Pressure, Concentration, and Unequal Conditions

Begin with the scene itself. Release a partition between unequal gas pressures and motion follows. Open a pathway between regions of unequal concentration and redistribution may follow. Place unequal temperatures in thermal contact and a thermal process may follow. It is easy to hurry past such an event because it is familiar. Yet familiarity often hides the number of conditions that must cooperate. The materials have properties. The surroundings have a history. A boundary permits some exchanges and suppresses others. The event looks simple only after all of those details have already been arranged.

Established science gives us the disciplined account. Each case has its own laws, transport coefficients, microscopic mechanisms, and boundary conditions. Similar language must not erase those differences. This account does not need a new vocabulary in order to remain valid. It tells us what to measure, which distinctions matter, and how the event can be compared with another event under specified conditions. Any deeper view must begin by respecting that achievement.

The Chavanian question begins one layer lower. The Chavanian view notices a shared architecture: a relational difference, an available pathway, a process of resolution, and a changed condition after the process. The word architecture directs attention away from an imagined travelling substance and toward the relations that make one response available while another remains blocked. This is an interpretation of the event, not a change to the event's measured description.

The same pattern becomes visible elsewhere. This architecture appears in porous media, physiology, chemical reactors, weather systems, electronic devices, and materials processing, although the measurable quantities differ. The comparison does not claim that the systems share an identical mechanism. It asks whether difference, constraint, pathway, and consequence provide a useful family resemblance across fields that often speak in separate dialects.

For teaching and early research, this distinction can be practical. Students can compare cases by asking the same structural questions while writing the established technical explanation separately for each case. A learner can state the observation first, the accepted scientific relation second, and the proposed interpretation third. That order reduces the risk that a helpful picture will be remembered as though it were the definition itself.

There is also a limit that should remain audible. A common architecture is not proof of a common fundamental mechanism. The comparison is a way of organizing inquiry. A conceptual account becomes scientifically stronger only when its terms can be constrained by measurement and when it accepts the possibility of failure. Until then, clarity is valuable, but clarity is not confirmation.

The section therefore returns to the chapter question. Different thermodynamic differences can participate in a comparable story without becoming the same physical quantity. In the Chavanian view, a change is not understood only by naming what changed. We also ask which difference mattered, which route was open, and what new condition the response created.

Pause once more over the ordinary details. Release a partition between unequal gas pressures and motion follows. Open a pathway between regions of unequal concentration and redistribution may follow. Place unequal temperatures in thermal contact and a thermal process may follow. Now imagine changing only one condition at a time: the material, the size, the surrounding medium, the duration of observation, or the character of the boundary. The visible event might remain recognizable, yet its pace and outcome could alter greatly. This is why a thermodynamic story becomes more accurate when it includes the setting instead of presenting the system as though it acted alone.

Language deserves its own small audit here. Different thermodynamic differences can participate in a comparable story without becoming the same physical quantity. In everyday speech we often turn a process into a thing. We say that heat enters, pressure pushes, or equilibrium arrives. Such expressions are efficient, and there is no need to ban them. The useful habit is to ask what the noun stands for in the actual account: a measured relation, an interaction, a redistribution, a constraint, or a shorthand for several of these together.

Maintained Difference

Consider what an observer would actually notice. A refrigerator keeps its interior cooler than the room. A cell maintains concentration differences across a membrane. Earth's atmosphere sustains temperature contrasts even while winds and radiation continually redistribute consequence. None of this observation tells us, by itself, which metaphor must be used underneath it. The event gives us readings, timings, boundaries, and consequences. The deeper story is something we add in order to make those relations intelligible.

Thermodynamics and the relevant specialist science already organize the event. Persistent non-equilibrium conditions require flows, work, reactions, external forcing, or other maintained processes. The wider system must be included in the accounting. These established descriptions are the common ground. They allow a teacher in one country, an engineer in another, and a researcher in a third field to speak about the same physical situation without depending on personal imagery.

Now introduce the Chavanian lens carefully. Delta can persist when the architecture continually recreates it or blocks the routes that would otherwise reduce it. Persistence is therefore an active system condition, not evidence that thermodynamic direction has failed. Instead of asking where an invisible content went, the lens asks how the system's condition, boundary, and available degrees of freedom shaped the response. The change of language is useful only if it produces greater discipline rather than a new ambiguity.

This way of speaking travels across disciplines with caution. Engineering control, biological regulation, ecological cycling, and planetary circulation all depend on maintained differences within open systems. The recurring architecture may be similar even when the mechanisms, scales, and measurable variables are very different. A membrane is not a piston, a cell is not an engine, and an atmosphere is not a sealed laboratory vessel. Analogy helps only when it does not erase those distinctions.

A teacher can turn this into a conversation. Ask what supplies the maintenance, where consequence is exported, and how the conclusion changes when the system boundary is widened. Ask which statement came from observation, which came from a tested model, and which came from interpretation. Students often discover that they understood the formal relation but had mixed several physical pictures together. Naming the layers gives them permission to separate those pictures without discarding the science.

The research boundary is equally important. Living maintenance should not be romanticized as an escape from thermodynamics. It is possible because the organism is open and participates in larger material and energetic processes. U E D does not gain experimental status merely because it offers a coherent vocabulary. Structured Vacuum Energy and Alignment Threshold Time require independent definitions, connections to existing measurements, and possible tests. The present book keeps that obligation visible.

What, then, should remain after this section? A difference may persist locally when a wider architecture continually pays the physical cost of maintaining it. The Chavanian view invites another question after the standard explanation is complete: what architecture made this particular thermodynamic response possible?

Let us return to the event without technical haste. A refrigerator keeps its interior cooler than the room. A cell maintains concentration differences across a membrane. Earth's atmosphere sustains temperature contrasts even while winds and radiation continually redistribute consequence. An observer could record what happens, but a fuller explanation would also ask what had to be present before it could happen. There had to be a particular arrangement, a particular contact with the surroundings, and a particular range of possible responses. The event is therefore not only a change in a quantity. It is a change made possible by a physical situation.

This is where ordinary grammar can quietly mislead us. A difference may persist locally when a wider architecture continually pays the physical cost of maintaining it. A convenient sentence may give an abstract quantity the role of an actor, as though the quantity itself travelled, commanded, or chose. Scientists usually understand the shorthand, but new learners may build their intuition around the grammar. Rephrasing the sentence in terms of conditions, interactions, and measurable change often reveals what the original expression had compressed.

Delta Has No Intention

A useful way to enter the topic is to slow the event down in imagination. We often say that heat seeks cold, pressure seeks release, or nature wants equilibrium. The language makes direction memorable but can make a relation sound like a decision-maker. At each moment, the system has a condition and a set of possible next steps. Some steps are easy, some require a threshold, and some are excluded by the way the system is built or connected.

The established account tells us how to describe that condition responsibly. Thermodynamic direction is expressed through established criteria and statistical behaviour under defined conditions. It does not require desire or purpose. It protects us from replacing measurement with intuition. Even in an equation-free conversation, the formal background matters because the words inherit their meaning from that disciplined practice.

From the Chavanian perspective, the event can be narrated as structured resolution. Delta is not an agent. It is the relevant difference in a configuration. Resolution occurs when a physically available process connects the conditions and the system evolves accordingly. Delta names the relevant difference. The architecture names the permitted and blocked routes. Resolution names the actual process. Consequence names the changed condition that remains after the process has occurred.

That sequence can illuminate another field. The same caution applies when discussing natural selection, chemical affinity, electrical potential, or neural activity. Direction does not automatically imply intention. Yet the illumination is conceptual rather than a claim that one universal mechanism has already been demonstrated. The value lies in asking comparable questions: What persists? What differs? What pathway is available? What must the surroundings supply or receive?

In a classroom or research group, the vocabulary can be used as a checking device. Replace the phrase the system wants with a description of the state difference, the available route, and the observed tendency. If the conversation begins to treat Delta as a force, architecture as a hidden machine, or resolution as a new measurable substance, the terms have drifted. Their purpose is to organize relations, not to populate nature with additional invisible objects.

The strongest caution is this. Removing agency from language does not remove causality or direction. It only prevents a relational condition from being mistaken for an actor. Established thermodynamics can stand without the Chavanian interpretation. The interpretation must therefore earn its place through explanatory usefulness now and through mathematical and experimental work later. It cannot demand immunity from the evidence that shaped the field.

The chapter's central movement appears again. Delta changes the set of possible outcomes without pushing like a hidden hand. Change becomes easier to discuss when we keep both the initiating difference and the available pathway in view. The answer is not hidden in either one alone.

Imagine watching the scene with the ability to stop it at several moments. We often say that heat seeks cold, pressure seeks release, or nature wants equilibrium. The language makes direction memorable but can make a relation sound like a decision-maker. At every pause, the system presents both a current condition and a limited range of next conditions. The range is shaped by material structure, contact, geometry, and surrounding constraints. Change is therefore neither a detached command nor a mysterious jump. It occurs through what the present arrangement can physically support.

A small change in wording can expose this structure. Delta changes the set of possible outcomes without pushing like a hidden hand. Instead of asking what substance moved, ask which relation changed. Instead of asking what quantity wanted to increase or decrease, ask which processes were permitted under the stated conditions. The older wording may still be retained for convenience, but the second wording makes its hidden assumptions easier to inspect. That is especially helpful when familiar metaphors begin to conflict with one another.

Difference After Resolution

Begin with the scene itself. When a hot object cools, the original temperature difference diminishes, but the surroundings become slightly warmer. When gas expands, its volume, pressure, temperature, and surroundings may enter new relations. It is easy to hurry past such an event because it is familiar. Yet familiarity often hides the number of conditions that must cooperate. The materials have properties. The surroundings have a history. A boundary permits some exchanges and suppresses others. The event looks simple only after all of those details have already been arranged.

Established science gives us the disciplined account. Thermodynamic processes connect states and can generate new gradients, entropy production, mechanical displacement, phase changes, and altered boundary conditions. This account does not need a new vocabulary in order to remain valid. It tells us what to measure, which distinctions matter, and how the event can be compared with another event under specified conditions. Any deeper view must begin by respecting that achievement.

The Chavanian question begins one layer lower. Resolution does not mean that difference vanishes from reality. It means that a particular Delta is transformed, redistributed, or replaced within a changed geometry. The word architecture directs attention away from an imagined travelling substance and toward the relations that make one response available while another remains blocked. This is an interpretation of the event, not a change to the event's measured description.

The same pattern becomes visible elsewhere. A chemical reaction changes composition and may create thermal and concentration differences. A storm reduces some atmospheric contrasts while rearranging moisture and momentum. The comparison does not claim that the systems share an identical mechanism. It asks whether difference, constraint, pathway, and consequence provide a useful family resemblance across fields that often speak in separate dialects.

For teaching and early research, this distinction can be practical. Have learners describe the system before and after the process, then list which differences diminished, which appeared, and which moved outside the chosen boundary. A learner can state the observation first, the accepted scientific relation second, and the proposed interpretation third. That order reduces the risk that a helpful picture will be remembered as though it were the definition itself.

There is also a limit that should remain audible. The language of new Delta is qualitative unless each relevant quantity is defined and measured. It should guide questions, not substitute for balances. A conceptual account becomes scientifically stronger only when its terms can be constrained by measurement and when it accepts the possibility of failure. Until then, clarity is valuable, but clarity is not confirmation.

The section therefore returns to the chapter question. Every resolution becomes part of the conditions from which later change begins. In the Chavanian view, a change is not understood only by naming what changed. We also ask which difference mattered, which route was open, and what new condition the response created.

Pause once more over the ordinary details. When a hot object cools, the original temperature difference diminishes, but the surroundings become slightly warmer. When gas expands, its volume, pressure, temperature, and surroundings may enter new relations. Now imagine changing only one condition at a time: the material, the size, the surrounding medium, the duration of observation, or the character of the boundary. The visible event might remain recognizable, yet its pace and outcome could alter greatly. This is why a thermodynamic story becomes more accurate when it includes the setting instead of presenting the system as though it acted alone.

Language deserves its own small audit here. Every resolution becomes part of the conditions from which later change begins. In everyday speech we often turn a process into a thing. We say that heat enters, pressure pushes, or equilibrium arrives. Such expressions are efficient, and there is no need to ban them. The useful habit is to ask what the noun stands for in the actual account: a measured relation, an interaction, a redistribution, a constraint, or a shorthand for several of these together.

Closing Reflection

Difference is neither a substance nor a command. It is a relation within a specified system. The Chavanian term Delta keeps that relation visible while reminding us that direction, pathway, scale, and boundary must still be supplied by the science of the particular case.

Delta begins the possibility of change. Architecture determines whether and how that possibility becomes an event.

Chapter 3
The Boundary Matters

What If the Boundary Helps Choose the Outcome?

The Wall Is Not Merely Where the Drawing Ends

Draw a box around a system in a textbook and the line may look passive. In a real experiment the boundary has thickness, composition, motion, permeability, roughness, thermal response, and history. It can conduct, insulate, deform, react, leak, select, or fail. The boundary is part of what happens.

The question in this chapter is not whether established thermodynamics works. It is what we notice when we place the idea of architecture beside walls, pistons, membranes, interfaces, and the definition of a system. The calculation and the measurement remain where science has placed them. Our task is to build a physical narrative that is clear enough for teaching, useful enough for interdisciplinary conversation, and modest enough to preserve the difference between interpretation and evidence.

We will move through familiar scenes, the established account, the Chavanian view, and the limits of that view. By the end, the phrase the boundary participates in the physical question should feel less like a slogan and more like a disciplined question that can accompany technical study without competing with it.

The Locked and Moving Piston

Begin with the scene itself. Heat a gas beneath a freely moving piston and expansion may share the response. Lock the piston and the same thermal intervention cannot use that mechanical route. It is easy to hurry past such an event because it is familiar. Yet familiarity often hides the number of conditions that must cooperate. The materials have properties. The surroundings have a history. A boundary permits some exchanges and suppresses others. The event looks simple only after all of those details have already been arranged.

Established science gives us the disciplined account. Established thermodynamics distinguishes constant-volume, constant-pressure, adiabatic, isothermal, reversible, and irreversible paths according to constraints and interactions. This account does not need a new vocabulary in order to remain valid. It tells us what to measure, which distinctions matter, and how the event can be compared with another event under specified conditions. Any deeper view must begin by respecting that achievement.

The Chavanian question begins one layer lower. The Chavanian view says the boundary architecture selects which modes of resolution are physically available. The gas does not receive an instruction called heat that carries one fixed outcome. The word architecture directs attention away from an imagined travelling substance and toward the relations that make one response available while another remains blocked.

This is an interpretation of the event, not a change to the event's measured description.

The same pattern becomes visible elsewhere. Flexible biological membranes, pressure relief devices, lungs, balloons, geological pores, and industrial reactors all show how boundary freedom changes system response. The comparison does not claim that the systems share an identical mechanism. It asks whether difference, constraint, pathway, and consequence provide a useful family resemblance across fields that often speak in separate dialects.

For teaching and early research, this distinction can be practical. Present the same initial gas in two imagined containers and ask students what the wall permits before asking them to predict the final state. A learner can state the observation first, the accepted scientific relation second, and the proposed interpretation third. That order reduces the risk that a helpful picture will be remembered as though it were the definition itself.

There is also a limit that should remain audible. The boundary does not choose consciously. The word choose refers to physical selection by constraint. A conceptual account becomes scientifically stronger only when its terms can be constrained by measurement and when it accepts the possibility of failure. Until then, clarity is valuable, but clarity is not confirmation.

The section therefore returns to the chapter question. The same broad input can appear differently because one boundary can move and another cannot. In the Chavanian view, a change is not understood only by naming what changed. We also ask which difference mattered, which route was open, and what new condition the response created.

Pause once more over the ordinary details. Heat a gas beneath a freely moving piston and expansion may share the response. Lock the piston and the same thermal intervention cannot use that mechanical route. Now imagine changing only one condition at a time: the material, the size, the surrounding medium, the duration of observation, or the character of the boundary. The visible event might remain recognizable, yet its pace and outcome could alter greatly. This is why a thermodynamic story becomes more accurate when it includes the setting instead of presenting the system as though it acted alone.

Language deserves its own small audit here. The same broad input can appear differently because one boundary can move and another cannot. In everyday speech we often turn a process into a thing. We say that heat enters, pressure pushes, or equilibrium arrives. Such expressions are efficient, and there is no need to ban them. The useful habit is to ask what the noun stands for in the actual account: a measured relation, an interaction, a redistribution, a constraint, or a shorthand for several of these together.

The Thermos and the Open Window

Consider what an observer would actually notice. A thermos slows exchange between its contents and the room. An open window connects indoor air with a changing outdoor environment. Neither boundary is perfect, but each reorganizes the pathways. None of this observation tells us, by itself, which metaphor must be used underneath it. The event gives us readings, timings, boundaries, and consequences. The deeper story is something we add in order to make those relations intelligible.

Thermodynamics and the relevant specialist science already organize the event. Insulation reduces modes and rates of heat transfer, while openings permit mass flow, convection, pressure adjustment, and moisture exchange. These established descriptions are the common ground. They allow a teacher in one country, an engineer in another, and a researcher in a third field to speak about the same physical situation without depending on personal imagery.

Now introduce the Chavanian lens carefully. Architecture is visible in what the boundary delays, allows, and couples. A Delta can remain unresolved longer when the easiest route is suppressed. Instead of asking where an invisible content went, the lens asks how the system's condition, boundary, and available degrees of freedom shaped the response. The change of language is useful only if it produces greater discipline rather than a new ambiguity.

This way of speaking travels across disciplines with caution. The same principle informs building envelopes, cryogenic vessels, incubators, environmental chambers, and spacecraft thermal control. The recurring architecture may be similar even when the mechanisms, scales, and measurable variables are very different. A membrane is not a piston, a cell is not an engine, and an atmosphere is not a sealed laboratory vessel. Analogy helps only when it does not erase those distinctions.

A teacher can turn this into a conversation. Ask students to replace the word closed with a description of what is actually restricted and over what timescale. Ask which statement came from observation, which came from a tested model, and which came from interpretation. Students often discover that they understood the formal relation but had mixed several physical pictures together. Naming the layers gives them permission to separate those pictures without discarding the science.

The research boundary is equally important. No practical boundary is simply open or closed in every respect. Matter, heat, radiation, and mechanical interaction may face different conditions. U E D does not gain experimental status merely because it offers a coherent vocabulary. Structured Vacuum Energy and Alignment Threshold Time require independent definitions, connections to existing measurements, and possible tests. The present book keeps that obligation visible.

What, then, should remain after this section? A boundary is best understood by the exchanges it permits and the timescales on which it permits them. The Chavanian view invites another question after the standard explanation is complete: what architecture made this particular thermodynamic response possible?

Let us return to the event without technical haste. A thermos slows exchange between its contents and the room. An open window connects indoor air with a changing outdoor environment. Neither boundary is perfect, but each reorganizes the pathways. An observer could record what happens, but a fuller explanation would also ask what had to be present before it could happen. There had to be a particular arrangement, a particular contact with the surroundings, and a particular range of possible responses. The event is therefore not only a change in a quantity. It is a change made possible by a physical situation.

This is where ordinary grammar can quietly mislead us. A boundary is best understood by the exchanges it permits and the timescales on which it permits them. A convenient sentence may give an abstract quantity the role of an actor, as though the quantity itself travelled, commanded, or chose. Scientists usually understand the shorthand, but new learners may build their intuition around the grammar. Rephrasing the sentence in terms of conditions, interactions, and measurable change often reveals what the original expression had compressed.

Selective Membranes

A useful way to enter the topic is to slow the event down in imagination. A membrane may allow water to pass while restricting ions, permit one gas more readily than another, or change its permeability in response to chemical and electrical conditions. At each moment, the system has a condition and a set of possible next steps. Some steps are easy, some require a threshold, and some are excluded by the way the system is built or connected.

The established account tells us how to describe that condition responsibly. Membrane transport depends on material structure, concentration, pressure, chemical potential, charge, channels, carriers, and active processes. It protects us from replacing measurement with intuition. Even in an equation-free conversation, the formal background matters because the words inherit their meaning from that disciplined practice.

From the Chavanian perspective, the event can be narrated as structured resolution. The Chavanian view calls this a selective resolution geometry. Several Deltas may coexist, but only certain routes are accessible to each species under the present conditions. Delta names the relevant difference. The architecture names the permitted and blocked routes. Resolution names the actual process. Consequence names the changed condition that remains after the process has occurred.

That sequence can illuminate another field. Cells, desalination systems, fuel cells, gas separation, dialysis, and geological barriers all depend on selective boundaries. Yet the illumination is conceptual rather than a claim that one universal mechanism has already been demonstrated. The value lies in asking comparable questions: What persists? What differs? What pathway is available? What must the surroundings supply or receive?

In a classroom or research group, the vocabulary can be used as a checking device. Have learners name the species, driving conditions, permitted route, blocked route, and source of any active transport. If the conversation begins to treat Delta as a force, architecture as a hidden machine, or resolution as a new measurable substance, the terms have drifted. Their purpose is to organize relations, not to populate nature with additional invisible objects.

The strongest caution is this. Selective resolution is an organizing phrase, not a replacement for transport theory or membrane kinetics. Established thermodynamics can stand without the Chavanian interpretation. The interpretation must therefore earn its place through explanatory usefulness now and through mathematical and experimental work later. It cannot demand immunity from the evidence that shaped the field.

The chapter's central movement appears again. A boundary can separate processes without separating reality into unrelated pieces. Change becomes easier to discuss when we keep both the initiating difference and the available pathway in view. The answer is not hidden in either one alone.

Imagine watching the scene with the ability to stop it at several moments. A membrane may allow water to pass while restricting ions, permit one gas more readily than another, or change its permeability in response to chemical and electrical conditions. At every pause, the system presents both a current condition and a limited range of next conditions. The range is shaped by material structure, contact, geometry, and surrounding constraints. Change is therefore neither a detached command nor a mysterious jump. It occurs through what the present arrangement can physically support.

A small change in wording can expose this structure. A boundary can separate processes without separating reality into unrelated pieces. Instead of asking what substance moved, ask which relation changed. Instead of asking what quantity wanted to increase or decrease, ask which processes were permitted under the stated conditions. The older wording may still be retained for convenience, but the second wording makes its hidden assumptions easier to inspect. That is especially helpful when familiar metaphors begin to conflict with one another.

Where the System Ends

Begin with the scene itself. The same refrigerator can be described as a cool compartment, as a machine plus compartment, or as a machine, kitchen, power station, and environment. Each boundary answers a different question. It is easy to hurry past such an event because it is familiar. Yet familiarity often hides the number of conditions that must cooperate. The materials have properties. The surroundings have a history. A boundary permits some exchanges and suppresses others. The event looks simple only after all of those details have already been arranged.

Established science gives us the disciplined account. Thermodynamic accounting depends on a clearly chosen system. Quantities entering or leaving one system may be internal processes within a wider system. This account does not need a new vocabulary in order to remain valid. It tells us what to measure, which distinctions matter, and how the event can be compared with another event under specified conditions. Any deeper view must begin by respecting that achievement.

The Chavanian question begins one layer lower. The Chavanian view treats boundaries as analytical and physical commitments. Widening the boundary can reveal where a maintained local Delta exports consequence. The word architecture directs attention away from an imagined travelling substance and toward the relations that make one response available while another remains blocked. This is an interpretation of the event, not a change to the event's measured description.

The same pattern becomes visible elsewhere. Life-cycle assessment, climate modelling, ecosystem analysis, and industrial efficiency all change when the system boundary is expanded. The comparison does not claim that the systems share an identical mechanism. It asks whether difference, constraint, pathway, and consequence provide a useful family resemblance across fields that often speak in separate dialects.

For teaching and early research, this distinction can be practical. Ask students to redraw the boundary three times and explain which exchanges disappear from the ledger because they have become internal. A learner can state the observation first, the accepted scientific relation second, and the proposed interpretation third. That order reduces the risk that a helpful picture will be remembered as though it were the definition itself.

There is also a limit that should remain audible. A wider boundary is not always better. The correct boundary is the one that matches the question while making assumptions explicit. A conceptual account becomes scientifically stronger only when its terms can be constrained by measurement and when it accepts the possibility of failure. Until then, clarity is valuable, but clarity is not confirmation.

The section therefore returns to the chapter question. Many apparent paradoxes shrink when the system boundary is chosen carefully and stated aloud. In the Chavanian view, a change is not understood only by naming what changed. We also ask which difference mattered, which route was open, and what new condition the response created.

Pause once more over the ordinary details. The same refrigerator can be described as a cool compartment, as a machine plus compartment, or as a machine, kitchen, power station, and environment. Each boundary answers a different question. Now imagine changing only one condition at a time: the material, the size, the surrounding medium, the duration of observation, or the character of the boundary. The visible event might remain recognizable, yet its pace and outcome could alter greatly. This is why a thermodynamic story becomes more accurate when it includes the setting instead of presenting the system as though it acted alone.

Language deserves its own small audit here. Many apparent paradoxes shrink when the system boundary is chosen carefully and stated aloud. In everyday speech we often turn a process into a thing. We say that heat enters, pressure pushes, or equilibrium arrives. Such expressions are efficient, and there is no need to ban them. The useful habit is to ask what the noun stands for in the actual account: a measured relation, an interaction, a redistribution, a constraint, or a shorthand for several of these together.

The Boundary as a Transitional Region

Consider what an observer would actually notice. At microscopic and mesoscopic scales, an interface may be a region with gradients, surface structure, adsorption, stress, and finite thickness rather than a perfectly sharp line. None of this observation tells us, by itself, which metaphor must be used underneath it. The event gives us readings, timings, boundaries, and consequences. The deeper story is something we add in order to make those relations intelligible.

Thermodynamics and the relevant specialist science already organize the event. Surface thermodynamics, interfacial science, fluid mechanics, and materials science describe how properties can differ near interfaces and how those regions affect transport and stability. These established descriptions are the common ground. They allow a teacher in one country, an engineer in another, and a researcher in a third field to speak about the same physical situation without depending on personal imagery.

Now introduce the Chavanian lens carefully. The Chavanian view finds this important because continuity does not require uniformity. A boundary can be a persistent transitional structure within a connected physical field. Instead of asking where an invisible content went, the lens asks how the system's condition, boundary, and available degrees of freedom shaped the response. The change of language is useful only if it produces greater discipline rather than a new ambiguity.

This way of speaking travels across disciplines with caution. Catalyst surfaces, cell membranes, phase fronts, cloud droplets, grain boundaries, and contact zones all make interfacial architecture physically consequential. The recurring architecture may be similar even when the mechanisms, scales, and measurable variables are very different. A membrane is not a piston, a cell is not an engine, and an atmosphere is not a sealed laboratory vessel. Analogy helps only when it does not erase those distinctions.

A teacher can turn this into a conversation. Move from the drawn line to the real material and ask what the line hides: thickness, composition, delay, roughness, storage, reaction, or deformation. Ask which statement came from observation, which came from a tested model, and which came from interpretation. Students often discover that they understood the formal relation but had mixed several physical pictures together. Naming the layers gives them permission to separate those pictures without discarding the science.

The research boundary is equally important. The U E D claim of deeper continuity goes beyond established interfacial science. Similar language must not be presented as evidence for Structured Vacuum Energy. U E D does not gain experimental status merely because it offers a coherent vocabulary. Structured Vacuum Energy and Alignment Threshold Time require independent definitions, connections to existing measurements, and possible tests. The present book keeps that obligation visible.

What, then, should remain after this section? The boundary is not merely where explanation stops. It is often where the decisive physics begins. The Chavanian view invites another question after the standard explanation is complete: what architecture made this particular thermodynamic response possible?

Let us return to the event without technical haste. At microscopic and mesoscopic scales, an interface may be a region with gradients, surface structure, adsorption, stress, and finite thickness rather than a perfectly sharp line. An observer could record what happens, but a fuller explanation would also ask what had to be present before it could happen. There had to be a particular arrangement, a particular contact with the surroundings, and a particular range of possible responses. The event is therefore not only a change in a quantity. It is a change made possible by a physical situation.

This is where ordinary grammar can quietly mislead us. The boundary is not merely where explanation stops. It is often where the decisive physics begins. A convenient sentence may give an abstract quantity the role of an actor, as though the quantity itself travelled, commanded, or chose. Scientists usually understand the shorthand, but new learners may build their intuition around the grammar. Rephrasing the sentence in terms of conditions, interactions, and measurable change often reveals what the original expression had compressed.

Closing Reflection

A boundary is a choice in analysis and a structure in the world. It determines what we count, what can cross, what must remain, and which responses are available. The Chavanian view places that architecture at the centre of the story while leaving the technical science of each boundary intact.

A thermodynamic process is shaped not only by the state of the system but by the freedoms and restrictions built into its boundary.

Interlude I

A Thermos, an Open Window, and a Question

Place warm tea in an ordinary cup and leave it on a table. Place the same amount of tea at the same initial temperature in a thermos. Put a third cup beside an open window on a cold day. The tea is similar. The room is broadly the same. Yet the three histories will not be the same.

The first cup exchanges thermal consequence through its walls, its open surface, the moving air, radiation, and evaporation. The thermos suppresses several of those routes and slows others. The cup by the window encounters stronger air movement and a different local environment. Nothing mysterious has been added to the tea. The architecture around the tea has changed.

This small scene contains the main movement of Part I. A thermal difference matters. The boundary participates. The available pathways determine the rate and visible form of the response. If we report only the initial and final temperatures, we may describe the states correctly while missing the physical story that connects them.

Established thermodynamics and heat-transfer science can analyse every version. They can define the system, identify mechanisms, measure material properties, and predict cooling behaviour. The Chavanian view does not compete with that work. It uses the scene to ask why the same broad thermal Delta becomes three different histories.

Now imagine closing the window halfway. Imagine placing a lid on the ordinary cup. Imagine stirring one cup while leaving another still. Imagine replacing the glass with metal. Each intervention changes the architecture. Some change the boundary material. Some change flow. Some change surface exposure. Some redistribute conditions inside the liquid.

This is why architecture should not be reduced to a wall. It includes the arrangement of the whole problem. It includes which parts touch, which parts move, which processes couple, how large the contact is, what the surroundings are doing, and how long the observation lasts.

For a teacher, the three cups are enough to begin a rich discussion without an equation. Students can predict qualitatively, justify their predictions, and then identify which formal topics would be needed for calculation. The conceptual question becomes a doorway into the technical one.

For a researcher, the scene is a reminder that boundary conditions are not administrative details placed after the governing model. They are part of the physical claim. A result obtained under one thermal geometry may not transfer to another simply because the material name and initial temperature are the same.

For the Chavanian view, the thermos offers another lesson. A Delta can remain relevant for longer when a pathway is restricted. Persistence does not mean that no difference exists. It may mean that the architecture prevents the difference from resolving quickly through the available modes.

Universal Energy Dynamics adds a further proposal: all three cups, all three boundaries, and all three environments exist within a deeper continuity. That proposal is not required to explain the cooling curves. It enters only when we ask what kind of physical reality lies beneath the connected event. The experiment supports thermodynamics. It does not, by itself, prove U E D.

The reading boundary can therefore be stated in one breath. The cups cool according to established thermal science. The Chavanian view organizes the event around Delta, boundary, pathway, and consequence. U E D proposes a continuous foundation beneath the event. The three levels can be discussed together as long as they are not merged.

The tea eventually becomes drinkable. That practical outcome is ordinary. The lesson is larger. Whenever two experiments begin with similar substances but reach different histories, look first at the architecture before inventing a new kind of content inside the system.

Part II

Beneath the Measured Quantities

→

Chapter 4

Reading Thermal Condition

What If Temperature Is a Reading of Condition?

The Thermometer Responds, but It Does Not Tell the Whole Story

A thermometer gives a number, and the number can be remarkably reliable. Yet the number is not the warmth we feel, the energy we imagine, the molecular history of the sample, or the total description of the system. Temperature is powerful partly because it says something precise without saying everything.

The question in this chapter is not whether established thermodynamics works. It is what we notice when we place the idea of architecture beside temperature as a thermodynamic coordinate and a condition of comparison. The calculation and the measurement remain where science has placed them. Our task is to build a physical narrative that is clear enough for teaching, useful enough for interdisciplinary conversation, and modest enough to preserve the difference between interpretation and evidence.

We will move through familiar scenes, the established account, the Chavanian view, and the limits of that view. By the end, the phrase temperature is a disciplined reading, not a substance inside matter should feel less like a slogan and more like a disciplined question that can accompany technical study without competing with it.

What a Thermometer Joins

Consider what an observer would actually notice. To measure temperature, the thermometer and the system must interact. After sufficient contact under suitable conditions, the instrument settles into a relation that supports a reproducible reading. None of this observation tells us, by itself, which metaphor must be used underneath it. The event gives us readings, timings, boundaries, and consequences. The deeper story is something we add in order to make those relations intelligible.

Thermodynamics and the relevant specialist science already organize the event. The Zeroth Law supports the transitive concept of thermal equilibrium and makes temperature a meaningful coordinate for comparing systems. These established descriptions are the common ground. They allow a teacher in one country, an engineer in another, and a researcher in a third field to speak about the same physical situation without depending on personal imagery.

Now introduce the Chavanian lens carefully. The Chavanian view describes measurement as a small resolution process between instrument and system. The reading records a relation achieved through a carefully designed boundary. Instead of asking where an invisible content went, the lens asks how the system's condition, boundary, and available degrees of freedom shaped the response. The change of language is useful only if it produces greater discipline rather than a new ambiguity.

This way of speaking travels across disciplines with caution. Clinical thermometers, remote radiation measurements, nanoscale probes, and atmospheric sensors each create different measurement architectures. The recurring architecture may be similar even when the mechanisms, scales, and measurable variables are very different. A membrane is not a piston, a cell is not an engine, and an atmosphere is not a sealed laboratory vessel. Analogy helps only when it does not erase those distinctions.

A teacher can turn this into a conversation. Ask what must happen physically before the displayed number becomes trustworthy and what assumptions connect the instrument's response to the system's temperature. Ask which statement came from observation, which came from a tested model, and which came from interpretation. Students often discover that they understood the formal relation but had mixed several physical pictures together. Naming the layers gives them permission to separate those pictures without discarding the science.

The research boundary is equally important. Measurement interaction is established science. Calling it resolution is interpretive and must not obscure calibration, uncertainty, or non-equilibrium effects. U E D does not gain experimental status merely because it offers a coherent vocabulary. Structured Vacuum Energy and Alignment Threshold Time require independent definitions, connections to existing measurements, and possible tests. The present book keeps that obligation visible.

What, then, should remain after this section? Temperature is not simply found inside an object. It is a state coordinate inferred through disciplined interaction and theory. The Chavanian view invites another question after the standard explanation is complete: what architecture made this particular thermodynamic response possible?

Let us return to the event without technical haste. To measure temperature, the thermometer and the system must interact. After sufficient contact under suitable conditions, the instrument settles into a relation that supports a reproducible reading. An observer could record what happens, but a fuller explanation would also ask what had to be present before it could happen. There had to be a particular arrangement, a particular contact with the surroundings, and a particular range of possible responses. The event is therefore not only a change in a quantity. It is a change made possible by a physical situation.

This is where ordinary grammar can quietly mislead us. Temperature is not simply found inside an object. It is a state coordinate inferred through disciplined interaction and theory. A convenient sentence may give an abstract quantity the role of an actor, as though the quantity itself travelled, commanded, or chose. Scientists usually understand the shorthand, but new learners may build their intuition around the grammar. Rephrasing the sentence in terms of conditions, interactions, and measurable change often reveals what the original expression had compressed.

Equal Temperature, Different Systems

A useful way to enter the topic is to slow the event down in imagination. A steel block, a container of water, and the surrounding air can share a temperature while differing greatly in mass, heat capacity, structure, and response to further heating. At each moment, the system has a condition and a set of possible next steps. Some steps are easy, some require a threshold, and some are excluded by the way the system is built or connected.

The established account tells us how to describe that condition responsibly. Thermal equilibrium in temperature does not make systems identical. Internal energy, phase, composition, pressure, and other state variables can remain different. It protects us from replacing measurement with intuition. Even in an equation-free conversation, the formal background matters because the words inherit their meaning from that disciplined practice.

From the Chavanian perspective, the event can be narrated as structured resolution. The Chavanian view calls equilibrium route-specific. The thermally resolvable Delta has diminished, while many persistent structural differences remain. Delta names the relevant difference. The architecture names the permitted and blocked routes. Resolution names the actual process. Consequence names the changed condition that remains after the process has occurred.

That sequence can illuminate another field. This matters in calorimetry, climate, medicine, food science, and materials processing, where equal temperature never implies equal thermal capacity or identical risk. Yet the illumination is conceptual rather than a claim that one universal mechanism has already been demonstrated. The value lies in asking comparable questions: What persists? What differs? What pathway is available? What must the surroundings supply or receive?

In a classroom or research group, the vocabulary can be used as a checking device. Invite students to list everything that can differ after thermal equilibrium has been reached. If the conversation begins to treat Delta as a force, architecture as a hidden machine, or resolution as a new measurable substance, the terms have drifted. Their purpose is to organize relations, not to populate nature with additional invisible objects.

The strongest caution is this. Route-specific is a conceptual phrase. The technical equilibrium conditions must still be stated using the established variables of the problem. Established thermodynamics can stand without the Chavanian interpretation. The interpretation must therefore earn its place through explanatory usefulness now and through mathematical and experimental work later. It cannot demand immunity from the evidence that shaped the field.

The chapter's central movement appears again. Equilibrium can remove one relevant difference while preserving the identities and capacities of the systems involved. Change becomes easier to discuss when we keep both the initiating difference and the available pathway in view. The answer is not hidden in either one alone.

Imagine watching the scene with the ability to stop it at several moments. A steel block, a container of water, and the surrounding air can share a temperature while differing greatly in mass, heat capacity, structure, and response to further heating. At every pause, the system presents both a current condition and a limited range of next conditions. The range is shaped by material structure, contact, geometry, and surrounding constraints. Change is therefore neither a detached command nor a mysterious jump. It occurs through what the present arrangement can physically support.

A small change in wording can expose this structure. Equilibrium can remove one relevant difference while preserving the identities and capacities of the systems involved. Instead of asking what substance moved, ask which relation changed. Instead of asking what quantity wanted to increase or decrease, ask which processes were permitted under the stated conditions. The older wording may still be retained for convenience, but the second wording makes its hidden assumptions easier to inspect. That is especially helpful when familiar metaphors begin to conflict with one another.

Local Temperature in a Changing World

Begin with the scene itself. A flame, an atmosphere, a turbulent fluid, or a living tissue can vary from place to place and moment to moment. A single temperature may describe only a small region or an average. It is easy to hurry past such an event because it is familiar. Yet familiarity often hides the number of conditions that must cooperate. The materials have properties. The surroundings have a history. A boundary permits some exchanges and suppresses others. The event looks simple only after all of those details have already been arranged.

Established science gives us the disciplined account. Local equilibrium approximations, transport theory, spatial resolution, response time, and measurement scale determine whether temperature is well defined and useful. This account does not need a new vocabulary in order to remain valid. It tells us what to measure, which distinctions matter, and how the event can be compared with another event under specified conditions. Any deeper view must begin by respecting that achievement.

The Chavanian question begins one layer lower. The Chavanian view asks which layer and timescale of the architecture the reading represents. Distributed systems may carry several unresolved thermal relations at once. The word architecture directs attention away from an imagined travelling substance and toward the relations that make one response available while another remains blocked. This is an interpretation of the event, not a change to the event's measured description.

The same pattern becomes visible elsewhere. Combustion, weather, microelectronics, biomedical imaging, and planetary science all depend on respecting spatial and temporal variation. The comparison does not claim that the systems share an identical mechanism. It asks whether difference, constraint, pathway, and consequence provide a useful family resemblance across fields that often speak in separate dialects.

For teaching and early research, this distinction can be practical. Ask whether the reported temperature belongs to a point, a region, an average, a surface, or an instrument with finite response time. A learner can state the observation first, the accepted scientific relation second, and the proposed interpretation third. That order reduces the risk that a helpful picture will be remembered as though it were the definition itself.

There is also a limit that should remain audible. The language of layers does not solve non-equilibrium thermodynamics. It only warns against treating every displayed number as a complete local state. A conceptual account becomes scientifically stronger only when its terms can be constrained by measurement and when it accepts the possibility of failure. Until then, clarity is valuable, but clarity is not confirmation.

The section therefore returns to the chapter question. A temperature reading has meaning within a scale, a location, a timescale, and a measurement architecture. In the Chavanian view, a change is not understood only by naming what changed. We also ask which difference mattered, which route was open, and what new condition the response created.

Pause once more over the ordinary details. A flame, an atmosphere, a turbulent fluid, or a living tissue can vary from place to place and moment to moment. A single temperature may describe only a small region or an average. Now imagine changing only one condition at a time: the material, the size, the surrounding medium, the duration of observation, or the character of the boundary. The visible event might remain recognizable, yet its pace and outcome could alter greatly. This is why a thermodynamic story becomes more accurate when it includes the setting instead of presenting the system as though it acted alone.

Language deserves its own small audit here. A temperature reading has meaning within a scale, a location, a timescale, and a measurement architecture. In everyday speech we often turn a process into a thing. We say that heat enters, pressure pushes, or equilibrium arrives. Such expressions are efficient, and there is no need to ban them. The useful habit is to ask what the noun stands for in the actual account: a measured relation, an interaction, a redistribution, a constraint, or a shorthand for several of these together.

Body Temperature and Regulation

Consider what an observer would actually notice. A living body can maintain a narrow internal temperature range while producing heat, exchanging matter, changing blood flow, sweating, shivering, and responding to its surroundings. None of this observation tells us, by itself, which metaphor must be used underneath it. The event gives us readings, timings, boundaries, and consequences. The deeper story is something we add in order to make those relations intelligible.

Thermodynamics and the relevant specialist science already organize the event. Thermoregulation involves metabolism, circulation, evaporation, conduction, convection, radiation, neural control, hormones, behaviour, and environmental conditions. These established descriptions are the common ground. They allow a teacher in one country, an engineer in another, and a researcher in a third field to speak about the same physical situation without depending on personal imagery.

Now introduce the Chavanian lens carefully. The Chavanian view treats the organism as an open architecture that actively manages thermal Deltas and redirects consequences through many coupled pathways. Instead of asking where an invisible content went, the lens asks how the system's condition, boundary, and available degrees of freedom shaped the response. The change of language is useful only if it produces greater discipline rather than a new ambiguity.

This way of speaking travels across disciplines with caution. The same perspective supports conversations about fever, exercise, animal adaptation, plant temperature, medical cooling, and thermal stress. The recurring architecture may be similar even when the mechanisms, scales, and measurable variables are very different. A membrane is not a piston, a cell is not an engine, and an atmosphere is not a sealed laboratory vessel. Analogy helps only when it does not erase those distinctions.

A teacher can turn this into a conversation. Begin with the body as an open system and ask what exchanges and regulatory actions make the maintained temperature possible. Ask which statement came from observation, which came from a tested model, and which came from interpretation. Students often discover that they understood the formal relation but had mixed several physical pictures together. Naming the layers gives them permission to separate those pictures without discarding the science.

The research boundary is equally important. Thermoregulation cannot be explained by thermodynamics alone. Physiology supplies mechanisms, control, and biological history. U E D does not gain experimental status merely because it offers a coherent vocabulary. Structured Vacuum Energy and Alignment Threshold Time require independent definitions, connections to existing measurements, and possible tests. The present book keeps that obligation visible.

What, then, should remain after this section? A maintained temperature is an achievement of an open system, not evidence that the system is thermodynamically isolated. The Chavanian view invites another question after the standard explanation is complete: what architecture made this particular thermodynamic response possible?

Let us return to the event without technical haste. A living body can maintain a narrow internal temperature range while producing heat, exchanging matter, changing blood flow, sweating, shivering, and responding to its surroundings. An observer could record what happens, but a fuller explanation would also ask what had to be present before it could happen. There had to be a particular arrangement, a particular contact with the surroundings, and a particular range of possible responses. The event is therefore not only a change in a quantity. It is a change made possible by a physical situation.

This is where ordinary grammar can quietly mislead us. A maintained temperature is an achievement of an open system, not evidence that the system is thermodynamically isolated. A convenient sentence may give an abstract quantity the role of an actor, as though the quantity itself travelled, commanded, or chose. Scientists usually understand the shorthand, but new learners may build their intuition around the grammar. Rephrasing the sentence in terms of conditions, interactions, and measurable change often reveals what the original expression had compressed.

Temperature Near the Limits

A useful way to enter the topic is to slow the event down in imagination. At very low temperatures, familiar modes of motion cease contributing in familiar ways, heat capacities change, quantum behaviour becomes unavoidable, and absolute zero remains unattainable by ordinary finite procedures. At each moment, the system has a condition and a set of possible next steps. Some steps are easy, some require a threshold, and some are excluded by the way the system is built or connected.

The established account tells us how to describe that condition responsibly. The Third Law, quantum statistics, low-temperature measurements, and material-specific behaviour govern this domain. It protects us from replacing measurement with intuition. Even in an equation-free conversation, the formal background matters because the words inherit their meaning from that disciplined practice.

From the Chavanian perspective, the event can be narrated as structured resolution. The Chavanian view asks whether diminishing thermal accessibility can be narrated as a narrowing architecture of available response, while preserving the established low-temperature results. Delta names the relevant difference. The architecture names the permitted and blocked routes. Resolution names the actual process. Consequence names the changed condition that remains after the process has occurred.

That sequence can illuminate another field. Cryogenics, superconductivity, quantum devices, low-temperature astronomy, and precision measurement all encounter the limits of classical thermal intuition. Yet the illumination is conceptual rather than a claim that one universal mechanism has already been demonstrated. The value lies in asking comparable questions: What persists? What differs? What pathway is available? What must the surroundings supply or receive?

In a classroom or research group, the vocabulary can be used as a checking device. Use the topic to show that temperature is not a simple measure of how much motion an object possesses in every regime. If the conversation begins to treat Delta as a force, architecture as a hidden machine, or resolution as a new measurable substance, the terms have drifted. Their purpose is to organize relations, not to populate nature with additional invisible objects.

The strongest caution is this. A Chavanian narrative cannot replace quantum statistical mechanics. Any future U E D account must recover its successful predictions before claiming deeper explanatory value. Established thermodynamics can stand without the Chavanian interpretation. The interpretation must therefore earn its place through explanatory usefulness now and through mathematical and experimental work later. It cannot demand immunity from the evidence that shaped the field.

The chapter's central movement appears again. The meaning of a thermodynamic coordinate is strengthened, not weakened, when we respect the regimes in which simple pictures fail. Change becomes easier to discuss when we keep both the initiating difference and the available pathway in view. The answer is not hidden in either one alone.

Imagine watching the scene with the ability to stop it at several moments. At very low temperatures, familiar modes of motion cease contributing in familiar ways, heat capacities change, quantum behaviour becomes unavoidable, and absolute zero remains unattainable by ordinary finite procedures. At every pause, the system presents both a current condition and a limited range of next conditions. The range is shaped by material structure, contact, geometry, and surrounding constraints. Change is therefore neither a detached command nor a mysterious jump. It occurs through what the present arrangement can physically support.

A small change in wording can expose this structure. The meaning of a thermodynamic coordinate is strengthened, not weakened, when we respect the regimes in which simple pictures fail. Instead of asking what substance moved, ask which relation changed. Instead of asking what quantity wanted to increase or decrease, ask which processes were permitted under the stated conditions. The older wording may still be retained for convenience, but the second wording makes its hidden assumptions easier to inspect. That is especially helpful when familiar metaphors begin to conflict with one another.

Closing Reflection

Temperature is among the most successful coordinates in science because it permits comparison across very different systems. The Chavanian view does not redefine it. It asks us to see the reading as part of an achieved relation within a measurement architecture and to remember that equal temperature leaves many other differences intact.

Temperature tells us something precise about thermal condition. It does not need to become a substance, a total energy content, or a complete portrait of the system.

Chapter 5
Pressure at the Edge

What If Pressure Is a Relationship at the Edge?

The Wall Feels the System

A gas in a cylinder presses on the wall. A liquid presses on the bottom of a vessel. The atmosphere presses on the surface of Earth and on every object within it. Pressure feels immediate because it can bend, lift, compress, and move things. Yet what exactly is present at the boundary?

The question in this chapter is not whether established thermodynamics works. It is what we notice when we place the idea of architecture beside pressure as a measurable state relation expressed through boundaries and fields. The calculation and the measurement remain where science has placed them. Our task is to build a physical narrative that is clear enough for teaching, useful enough for interdisciplinary conversation, and modest enough to preserve the difference between interpretation and evidence.

We will move through familiar scenes, the established account, the Chavanian view, and the limits of that view. By the end, the phrase pressure is not a stored push waiting to escape should feel less like a slogan and more like a disciplined question that can accompany technical study without competing with it.

Gas Against a Wall

A useful way to enter the topic is to slow the event down in imagination. A gas appears calm at human scale while its microscopic constituents interact continually. The wall experiences a sustained macroscopic pressure that can be measured and used to predict mechanical response. At each moment, the system has a condition and a set of possible next steps. Some steps are easy, some require a threshold, and some are excluded by the way the system is built or connected.

The established account tells us how to describe that condition responsibly. Kinetic theory connects gas pressure with microscopic momentum transfer, while thermodynamics treats pressure as a state variable and generalized force paired with volume change. It protects us from replacing measurement with intuition. Even in an equation-free conversation, the formal background matters because the words inherit their meaning from that disciplined practice.

From the Chavanian perspective, the event can be narrated as structured resolution. The Chavanian view describes pressure as a boundary manifestation of the system's constrained internal condition within a persistent geometry. Delta names the relevant difference. The architecture names the permitted and blocked routes. Resolution names the actual process. Consequence names the changed condition that remains after the process has occurred.

That sequence can illuminate another field. The same boundary emphasis matters in vacuum systems, pneumatic tools, lungs, tires, and gas storage, although material response and flow may become essential. Yet the illumination is conceptual rather than a claim that one universal mechanism has already been demonstrated. The value lies in asking comparable questions: What persists? What differs? What pathway is available? What must the surroundings supply or receive?

In a classroom or research group, the vocabulary can be used as a checking device. Ask whether pressure belongs only to the wall, only to the gas, or to the relation expressed where the gas and wall meet. If the conversation begins to treat Delta as a force, architecture as a hidden machine, or resolution as a new measurable substance, the terms have drifted. Their purpose is to organize relations, not to populate nature with additional invisible objects.

The strongest caution is this. Boundary manifestation is an interpretation and must not be used to dismiss kinetic theory or continuum mechanics. Established thermodynamics can stand without the Chavanian interpretation. The interpretation must therefore earn its place through explanatory usefulness now and through mathematical and experimental work later. It cannot demand immunity from the evidence that shaped the field.

The chapter's central movement appears again. Pressure is measurable at a boundary because the system and boundary participate in one constrained physical relation. Change becomes easier to discuss when we keep both the initiating difference and the available pathway in view. The answer is not hidden in either one alone.

Imagine watching the scene with the ability to stop it at several moments. A gas appears calm at human scale while its microscopic constituents interact continually. The wall experiences a sustained macroscopic pressure that can be measured and used to predict mechanical response. At every pause, the system presents both a current condition and a limited range of next conditions. The range is shaped by material structure, contact, geometry, and surrounding constraints. Change is therefore neither a detached command nor a mysterious jump. It occurs through what the present arrangement can physically support.

A small change in wording can expose this structure. Pressure is measurable at a boundary because the system and boundary participate in one constrained physical relation. Instead of asking what substance moved, ask which relation changed. Instead of asking what quantity wanted to increase or decrease, ask which processes were permitted under the stated conditions. The older wording may still be retained for convenience, but the second wording makes its hidden assumptions easier to inspect. That is especially helpful when familiar metaphors begin to conflict with one another.

Pressure Inside a Fluid

Begin with the scene itself. In a fluid under gravity, pressure changes with depth. A diver, a dam wall, and a sensor at the bottom encounter conditions different from those near the surface. It is easy to hurry past such an event because it is familiar. Yet familiarity often hides the number of conditions that must cooperate. The materials have properties. The surroundings have a history. A boundary permits some exchanges and suppresses others. The event looks simple only after all of those details have already been arranged.

Established science gives us the disciplined account. Hydrostatic pressure reflects the weight and mechanical equilibrium of the fluid under gravity, while flowing fluids require a fuller dynamical account. This account does not need a new vocabulary in order to remain valid. It tells us what to measure, which distinctions matter, and how the event can be compared with another event under specified conditions. Any deeper view must begin by respecting that achievement.

The Chavanian question begins one layer lower. The Chavanian view asks how the persistent field of conditions is distributed through the fluid and expressed locally at boundaries and probes. The word architecture directs attention away from an imagined travelling substance and toward the relations that make one response available while another remains blocked. This is an interpretation of the event, not a change to the event's measured description.

The same pattern becomes visible elsewhere. Oceanography, blood circulation, groundwater, hydraulic systems, and planetary interiors all depend on spatial pressure structure. The comparison does not claim that the systems share an identical mechanism. It asks whether difference, constraint, pathway, and consequence provide a useful family resemblance across fields that often speak in separate dialects.

For teaching and early research, this distinction can be practical. Use several imagined probes at different depths and ask what each measures, what remains the same, and which forces belong to the wider system. A learner can state the observation first, the accepted scientific relation second, and the proposed interpretation third. That order reduces the risk that a helpful picture will be remembered as though it were the definition itself.

There is also a limit that should remain audible. Pressure should not be treated as a substance flowing from high to low. Fluid motion emerges from pressure gradients together with inertia, viscosity, gravity, geometry, and other conditions. A conceptual account becomes scientifically stronger only when its terms can be constrained by measurement and when it accepts the possibility of failure. Until then, clarity is valuable, but clarity is not confirmation.

The section therefore returns to the chapter question. Pressure is a local thermodynamic and mechanical relation within a larger field of conditions. In the Chavanian view, a change is not understood only by naming what changed. We also ask which difference mattered, which route was open, and what new condition the response created.

Pause once more over the ordinary details. In a fluid under gravity, pressure changes with depth. A diver, a dam wall, and a sensor at the bottom encounter conditions different from those near the surface. Now imagine changing only one condition at a time: the material, the size, the surrounding medium, the duration of observation, or the character of the boundary. The visible event might remain recognizable, yet its pace and outcome could alter greatly. This is why a thermodynamic story becomes more accurate when it includes the setting instead of presenting the system as though it acted alone.

Language deserves its own small audit here. Pressure is a local thermodynamic and mechanical relation within a larger field of conditions. In everyday speech we often turn a process into a thing. We say that heat enters, pressure pushes, or equilibrium arrives. Such expressions are efficient, and there is no need to ban them. The useful habit is to ask what the noun stands for in the actual account: a measured relation, an interaction, a redistribution, a constraint, or a shorthand for several of these together.

Pressure and Confinement

Consider what an observer would actually notice. Compressing a gas changes the relation between its internal state and the container. Releasing a valve can create flow, cooling, sound, turbulence, and new pressure differences. None of this observation tells us, by itself, which metaphor must be used underneath it. The event gives us readings, timings, boundaries, and consequences. The deeper story is something we add in order to make those relations intelligible.

Thermodynamics and the relevant specialist science already organize the event. The response depends on the process path, equation of state, heat exchange, flow regime, valve geometry, and properties of the gas. These established descriptions are the common ground. They allow a teacher in one country, an engineer in another, and a researcher in a third field to speak about the same physical situation without depending on personal imagery.

Now introduce the Chavanian lens carefully. The Chavanian view emphasizes imposed geometrical Delta and the route through which confinement changes. The pressure does not escape as a stored object; the connected system evolves. Instead of asking where an invisible content went, the lens asks how the system's condition, boundary, and available degrees of freedom shaped the response. The change of language is useful only if it produces greater discipline rather than a new ambiguity.

This way of speaking travels across disciplines with caution. Aerosols, propulsion, refrigeration, compressed-air storage, volcanology, and respiratory mechanics all involve confinement and release. The recurring architecture may be similar even when the mechanisms, scales, and measurable variables are very different. A membrane is not a piston, a cell is not an engine, and an atmosphere is not a sealed laboratory vessel. Analogy helps only when it does not erase those distinctions.

A teacher can turn this into a conversation. Ask students to narrate the release as a sequence of changed boundaries and states rather than as pressure leaving the tank. Ask which statement came from observation, which came from a tested model, and which came from interpretation. Students often discover that they understood the formal relation but had mixed several physical pictures together. Naming the layers gives them permission to separate those pictures without discarding the science.

The research boundary is equally important. Conversational language must not erase hazards. Rapid decompression is a complex irreversible process that requires technical analysis. U E D does not gain experimental status merely because it offers a coherent vocabulary. Structured Vacuum Energy and Alignment Threshold Time require independent definitions, connections to existing measurements, and possible tests. The present book keeps that obligation visible.

What, then, should remain after this section? Release changes the geometry that sustained the prior state, allowing new motion and redistribution to occur. The Chavanian view invites another question after the standard explanation is complete: what architecture made this particular thermodynamic response possible?

Let us return to the event without technical haste. Compressing a gas changes the relation between its internal state and the container. Releasing a valve can create flow, cooling, sound, turbulence, and new pressure differences. An observer could record what happens, but a fuller explanation would also ask what had to be present before it could happen. There had to be a particular arrangement, a particular contact with the surroundings, and a particular range of possible responses. The event is therefore not only a change in a quantity. It is a change made possible by a physical situation.

This is where ordinary grammar can quietly mislead us. Release changes the geometry that sustained the prior state, allowing new motion and redistribution to occur. A convenient sentence may give an abstract quantity the role of an actor, as though the quantity itself travelled, commanded, or chose. Scientists usually understand the shorthand, but new learners may build their intuition around the grammar. Rephrasing the sentence in terms of conditions, interactions, and measurable change often reveals what the original expression had compressed.

Atmospheric Pressure and Weather

A useful way to enter the topic is to slow the event down in imagination. Atmospheric pressure varies across height, location, and time. Those variations participate in winds and weather, but no air parcel responds to pressure alone. At each moment, the system has a condition and a set of possible next steps. Some steps are easy, some require a threshold, and some are excluded by the way the system is built or connected.

The established account tells us how to describe that condition responsibly. Atmospheric motion involves pressure gradients, gravity, rotation, temperature, moisture, radiation, topography, turbulence, and phase changes. It protects us from replacing measurement with intuition. Even in an equation-free conversation, the formal background matters because the words inherit their meaning from that disciplined practice.

From the Chavanian perspective, the event can be narrated as structured resolution. The Chavanian view sees weather as a layered resolution architecture in which several Deltas interact and continually recreate one another. Delta names the relevant difference. The architecture names the permitted and blocked routes. Resolution names the actual process. Consequence names the changed condition that remains after the process has occurred.

That sequence can illuminate another field. The picture connects laboratory thermodynamics to geophysical fluid dynamics while reminding us that scale and open boundaries transform the problem. Yet the illumination is conceptual rather than a claim that one universal mechanism has already been demonstrated. The value lies in asking comparable questions: What persists? What differs? What pathway is available? What must the surroundings supply or receive?

In a classroom or research group, the vocabulary can be used as a checking device. Use a weather map to ask which differences are shown, which are hidden, and why a single gradient cannot predict the complete evolution. If the conversation begins to treat Delta as a force, architecture as a hidden machine, or resolution as a new measurable substance, the terms have drifted. Their purpose is to organize relations, not to populate nature with additional invisible objects.

The strongest caution is this. A broad architecture is not a weather model. Quantitative atmospheric prediction requires established dynamical equations, observations, and numerical methods. Established thermodynamics can stand without the Chavanian interpretation. The interpretation must therefore earn its place through explanatory usefulness now and through mathematical and experimental work later. It cannot demand immunity from the evidence that shaped the field.

The chapter's central movement appears again. Pressure contributes to atmospheric change within a coupled architecture rather than acting as an isolated cause. Change becomes easier to discuss when we keep both the initiating difference and the available pathway in view. The answer is not hidden in either one alone.

Imagine watching the scene with the ability to stop it at several moments. Atmospheric pressure varies across height, location, and time. Those variations participate in winds and weather, but no air parcel responds to pressure alone. At every pause, the system presents both a current condition and a limited range of next conditions. The range is shaped by material structure, contact, geometry, and surrounding constraints. Change is therefore neither a detached command nor a mysterious jump. It occurs through what the present arrangement can physically support.

A small change in wording can expose this structure. Pressure contributes to atmospheric change within a coupled architecture rather than acting as an isolated cause. Instead of asking what substance moved, ask which relation changed. Instead of asking what quantity wanted to increase or decrease, ask which processes were permitted under the stated conditions. The older wording may still be retained for convenience, but the second wording makes its hidden assumptions easier to inspect. That is especially helpful when familiar metaphors begin to conflict with one another.

When Pressure Becomes a Character

Begin with the scene itself. We say pressure builds, searches for release, pushes its way out, or forces a system to act. The language is vivid because pressure has obvious mechanical consequences. It is easy to hurry past such an event because it is familiar. Yet familiarity often hides the number of conditions that must cooperate. The materials have properties. The surroundings have a history. A boundary permits some exchanges and suppresses others. The event looks simple only after all of those details have already been arranged.

Established science gives us the disciplined account. Pressure enters precise constitutive, mechanical, and thermodynamic relations. Its role depends on the system and process being described. This account does not need a new vocabulary in order to remain valid. It tells us what to measure, which distinctions matter, and how the event can be compared with another event under specified conditions. Any deeper view must begin by respecting that achievement.

The Chavanian question begins one layer lower. The Chavanian view removes the character and keeps the relation. A pressure Delta alters the available mechanical responses of a connected geometry. The word architecture directs attention away from an imagined travelling substance and toward the relations that make one response available while another remains blocked. This is an interpretation of the event, not a change to the event's measured description.

The same pattern becomes visible elsewhere. The same linguistic caution applies to stress in solids, osmotic pressure in solutions, radiation pressure, and pressure-like variables in specialized theories. The comparison does not claim that the systems share an identical mechanism. It asks whether difference, constraint, pathway, and consequence provide a useful family resemblance across fields that often speak in separate dialects.

For teaching and early research, this distinction can be practical. Ask learners to replace the phrase pressure wants with a statement about measured differences, forces on boundaries, and permitted motion. A learner can state the observation first, the accepted scientific relation second, and the proposed interpretation third. That order reduces the risk that a helpful picture will be remembered as though it were the definition itself.

There is also a limit that should remain audible. Not every quantity called pressure has the same microscopic origin. The shared word should not hide domain-specific definitions. A conceptual account becomes scientifically stronger only when its terms can be constrained by measurement and when it accepts the possibility of failure. Until then, clarity is valuable, but clarity is not confirmation.

The section therefore returns to the chapter question. Pressure has consequence without intention. It describes a condition from which motion or deformation may become available. In the Chavanian view, a change is not understood only by naming what changed. We also ask which difference mattered, which route was open, and what new condition the response created.

Pause once more over the ordinary details. We say pressure builds, searches for release, pushes its way out, or forces a system to act. The language is vivid because pressure has obvious mechanical consequences. Now imagine changing only one condition at a time: the material, the size, the surrounding medium, the duration of observation, or the character of the boundary. The visible event might remain recognizable, yet its pace and outcome could alter greatly. This is why a thermodynamic story becomes more accurate when it includes the setting instead of presenting the system as though it acted alone.

Language deserves its own small audit here. Pressure has consequence without intention. It describes a condition from which motion or deformation may become available. In everyday speech we often turn a process into a thing. We say that heat enters, pressure pushes, or equilibrium arrives. Such expressions are efficient, and there is no need to ban them. The useful habit is to ask what the noun stands for in the actual account: a measured relation, an interaction, a redistribution, a constraint, or a shorthand for several of these together.

Closing Reflection

Pressure becomes clearer when it is allowed to remain a measured relation instead of becoming an invisible content. It can be described through kinetic theory, continuum mechanics, fields, and thermodynamic state relations. The Chavanian contribution is to keep the boundary and geometry visible in the physical story.

Pressure does not wait inside a system as a substance. It is expressed through the constrained relation between a system, its state, and the boundaries it encounters.

Chapter 6
Equilibrium Without Stillness

What If Equilibrium Is an Achievement?

Stillness Is Not the Same as Completion

Two bodies reach the same temperature. A reaction mixture reaches equilibrium. A gas settles into a stable macroscopic condition. We often picture equilibrium as stillness, yet microscopic activity can continue and persistent structure can remain. What has actually been achieved?

The question in this chapter is not whether established thermodynamics works. It is what we notice when we place the idea of architecture beside thermal, mechanical, chemical, and dynamic forms of equilibrium. The calculation and the measurement remain where science has placed them. Our task is to build a physical narrative that is clear enough for teaching, useful enough for interdisciplinary conversation, and modest enough to preserve the difference between interpretation and evidence.

We will move through familiar scenes, the established account, the Chavanian view, and the limits of that view. By the end, the phrase equilibrium can preserve structure after a particular route of change is exhausted should feel less like a slogan and more like a disciplined question that can accompany technical study without competing with it.

Thermal Equilibrium Without Sameness

Begin with the scene itself. A metal block and a container of water can reach the same temperature while retaining different compositions, masses, structures, and capacities. It is easy to hurry past such an event because it is familiar. Yet familiarity often hides the number of conditions that must cooperate. The materials have properties. The surroundings have a history. A boundary permits some exchanges and suppresses others. The event looks simple only after all of those details have already been arranged.

Established science gives us the disciplined account. Thermal equilibrium removes a net thermally driven exchange under the stated conditions and supports a common temperature. This account does not need a new vocabulary in order to remain valid. It tells us what to measure, which distinctions matter, and how the event can be compared with another event under specified conditions. Any deeper view must begin by respecting that achievement.

The Chavanian question begins one layer lower. The Chavanian view says the relevant thermal Delta is no longer resolvable through the available contact, even though many persistent differences remain. The word architecture directs attention away from an imagined travelling substance and toward the relations that make one response available while another remains blocked. This is an interpretation of the event, not a change to the event's measured description.

The same pattern becomes visible elsewhere. The distinction helps in calorimetry, climate discussions, physiology, and materials science, where equal temperature never means identical state. The comparison does not claim that the systems share an identical mechanism. It asks whether difference, constraint, pathway, and consequence provide a useful family resemblance across fields that often speak in separate dialects.

For teaching and early research, this distinction can be practical. Ask students to describe equilibrium first in the relevant variable and then list every form of difference that survives. A learner can state the observation first, the accepted scientific relation second, and the proposed interpretation third. That order reduces the risk that a helpful picture will be remembered as though it were the definition itself.

There is also a limit that should remain audible. Equilibrium conditions are technically defined. The phrase exhausted route is an interpretation of those conditions. A conceptual account becomes scientifically stronger only when its terms can be constrained by measurement and when it accepts the possibility of failure. Until then, clarity is valuable, but clarity is not confirmation.

The section therefore returns to the chapter question. Equilibrium can be specific to a mode of exchange while the systems remain richly different. In the Chavanian view, a change is not understood only by naming what changed. We also ask which difference mattered, which route was open, and what new condition the response created.

Pause once more over the ordinary details. A metal block and a container of water can reach the same temperature while retaining different compositions, masses, structures, and capacities. Now imagine changing only one condition at a time: the material, the size, the surrounding medium, the duration of observation, or the character of the boundary. The visible event might remain recognizable, yet its pace and outcome could alter greatly. This is why a thermodynamic story becomes more accurate when it includes the setting instead of presenting the system as though it acted alone.

Language deserves its own small audit here. Equilibrium can be specific to a mode of exchange while the systems remain richly different. In everyday speech we often turn a process into a thing. We say that heat enters, pressure pushes, or equilibrium arrives. Such expressions are efficient, and there is no need to ban them. The useful habit is to ask what the noun stands for in the actual account: a measured relation, an interaction, a redistribution, a constraint, or a shorthand for several of these together.

Microscopic Activity, Macroscopic Stability

Consider what an observer would actually notice. A gas at equilibrium is not a collection of frozen particles. A liquid surface continues fluctuating. Molecules continue colliding, rotating, and exchanging momentum. None of this observation tells us, by itself, which metaphor must be used underneath it. The event gives us readings, timings, boundaries, and consequences. The deeper story is something we add in order to make those relations intelligible.

Thermodynamics and the relevant specialist science already organize the event. Statistical mechanics connects persistent macroscopic state variables with distributions of microscopic configurations and ongoing fluctuations. These established descriptions are the common ground. They allow a teacher in one country, an engineer in another, and a researcher in a third field to speak about the same physical situation without depending on personal imagery.

Now introduce the Chavanian lens carefully. The Chavanian view distinguishes activity from unresolved macroscopic Delta. A system can remain dynamically active while its larger thermodynamic condition is stable. Instead of asking where an invisible content went, the lens asks how the system's condition, boundary, and available degrees of freedom shaped the response. The change of language is useful only if it produces greater discipline rather than a new ambiguity.

This way of speaking travels across disciplines with caution. This distinction appears in chemical exchange, biological turnover, market metaphors, and ecological steady behaviour, although only the physical cases belong directly to thermodynamics. The recurring architecture may be similar even when the mechanisms, scales, and measurable variables are very different. A membrane is not a piston, a cell is not an engine, and an atmosphere is not a sealed laboratory vessel. Analogy helps only when it does not erase those distinctions.

A teacher can turn this into a conversation. Challenge the image of equilibrium as frozen stillness and ask what continues to happen without altering the measured macroscopic state. Ask which statement came from observation, which came from a tested model, and which came from interpretation. Students often discover that they understood the formal relation but had mixed several physical pictures together. Naming the layers gives them permission to separate those pictures without discarding the science.

The research boundary is equally important. The Chavanian language cannot replace statistical distributions or fluctuation theory. U E D does not gain experimental status merely because it offers a coherent vocabulary. Structured Vacuum Energy and Alignment Threshold Time require independent definitions, connections to existing measurements, and possible tests. The present book keeps that obligation visible.

What, then, should remain after this section? Macroscopic persistence can emerge from continuous microscopic activity without contradiction. The Chavanian view invites another question after the standard explanation is complete: what architecture made this particular thermodynamic response possible?

Let us return to the event without technical haste. A gas at equilibrium is not a collection of frozen particles. A liquid surface continues fluctuating. Molecules continue colliding, rotating, and exchanging momentum. An observer could record what happens, but a fuller explanation would also ask what had to be present before it could happen. There had to be a particular arrangement, a particular contact with the surroundings, and a particular range of possible responses. The event is therefore not only a change in a quantity. It is a change made possible by a physical situation.

This is where ordinary grammar can quietly mislead us. Macroscopic persistence can emerge from continuous microscopic activity without contradiction. A convenient sentence may give an abstract quantity the role of an actor, as though the quantity itself travelled, commanded, or chose. Scientists usually understand the shorthand, but new learners may build their intuition around the grammar. Rephrasing the sentence in terms of conditions, interactions, and measurable change often reveals what the original expression had compressed.

Chemical Equilibrium as Continuing Exchange

A useful way to enter the topic is to slow the event down in imagination. In a reversible reaction at equilibrium, forward and reverse molecular events can continue while macroscopic composition remains stable under fixed conditions. At each moment, the system has a condition and a set of possible next steps. Some steps are easy, some require a threshold, and some are excluded by the way the system is built or connected.

The established account tells us how to describe that condition responsibly. Chemical thermodynamics characterizes equilibrium through established potentials, activities, composition, temperature, and pressure. It protects us from replacing measurement with intuition. Even in an equation-free conversation, the formal background matters because the words inherit their meaning from that disciplined practice.

From the Chavanian perspective, the event can be narrated as structured resolution. The Chavanian view interprets the stable composition as a condition in which the available reaction architecture no longer produces net macroscopic change. Delta names the relevant difference. The architecture names the permitted and blocked routes. Resolution names the actual process. Consequence names the changed condition that remains after the process has occurred.

That sequence can illuminate another field. The idea is relevant to catalysis, atmospheric chemistry, biochemistry, electrochemistry, and phase equilibria. Yet the illumination is conceptual rather than a claim that one universal mechanism has already been demonstrated. The value lies in asking comparable questions: What persists? What differs? What pathway is available? What must the surroundings supply or receive?

In a classroom or research group, the vocabulary can be used as a checking device. Ask students why a catalyst can change the rate of approach without changing the equilibrium condition under the same thermodynamic constraints. If the conversation begins to treat Delta as a force, architecture as a hidden machine, or resolution as a new measurable substance, the terms have drifted. Their purpose is to organize relations, not to populate nature with additional invisible objects.

The strongest caution is this. Resolution language must not be confused with reaction mechanism or kinetics. Those require independent chemical description. Established thermodynamics can stand without the Chavanian interpretation. The interpretation must therefore earn its place through explanatory usefulness now and through mathematical and experimental work later. It cannot demand immunity from the evidence that shaped the field.

The chapter's central movement appears again. Equilibrium can mean no net macroscopic change even while events continue in both directions. Change becomes easier to discuss when we keep both the initiating difference and the available pathway in view. The answer is not hidden in either one alone.

Imagine watching the scene with the ability to stop it at several moments. In a reversible reaction at equilibrium, forward and reverse molecular events can continue while macroscopic composition remains stable under fixed conditions. At every pause, the system presents both a current condition and a limited range of next conditions. The range is shaped by material structure, contact, geometry, and surrounding constraints. Change is therefore neither a detached command nor a mysterious jump. It occurs through what the present arrangement can physically support.

A small change in wording can expose this structure. Equilibrium can mean no net macroscopic change even while events continue in both directions. Instead of asking what substance moved, ask which relation changed. Instead of asking what quantity wanted to increase or decrease, ask which processes were permitted under the stated conditions. The older wording may still be retained for convenience, but the second wording makes its hidden assumptions easier to inspect. That is especially helpful when familiar metaphors begin to conflict with one another.

Steady State Is Not Equilibrium

Begin with the scene itself. A flowing river can maintain a roughly stable level while water continually enters and leaves. A living cell can maintain internal concentrations through ongoing transport and reaction. It is easy to hurry past such an event because it is familiar. Yet familiarity often hides the number of conditions that must cooperate. The materials have properties. The surroundings have a history. A boundary permits some exchanges and suppresses others. The event looks simple only after all of those details have already been arranged.

Established science gives us the disciplined account. A steady state can preserve macroscopic variables through balanced flows and production, whereas equilibrium requires the absence of relevant net thermodynamic driving under the specified constraints. This account does not need a new vocabulary in order to remain valid. It tells us what to measure, which distinctions matter, and how the event can be compared with another event under specified conditions. Any deeper view must begin by respecting that achievement.

The Chavanian question begins one layer lower. The Chavanian view describes steady persistence as maintained architecture. Deltas remain and are continually managed rather than exhausted. The word architecture directs attention away from an imagined travelling substance and toward the relations that make one response available while another remains blocked. This is an interpretation of the event, not a change to the event's measured description.

The same pattern becomes visible elsewhere. Reactors, organisms, climate systems, electrical circuits, and industrial processes often operate in non-equilibrium steady states. The comparison does not claim that the systems share an identical mechanism. It asks whether difference, constraint, pathway, and consequence provide a useful family resemblance across fields that often speak in separate dialects.

For teaching and early research, this distinction can be practical. Ask what must continue entering, leaving, or being produced for the apparently stable condition to persist. A learner can state the observation first, the accepted scientific relation second, and the proposed interpretation third. That order reduces the risk that a helpful picture will be remembered as though it were the definition itself.

There is also a limit that should remain audible. Calling both conditions balanced can hide the decisive distinction between maintained throughput and equilibrium. A conceptual account becomes scientifically stronger only when its terms can be constrained by measurement and when it accepts the possibility of failure. Until then, clarity is valuable, but clarity is not confirmation.

The section therefore returns to the chapter question. A stable reading may represent achieved equilibrium or sustained non-equilibrium, and the system history reveals which one it is. In the Chavanian view, a change is not understood only by naming what changed. We also ask which difference mattered, which route was open, and what new condition the response created.

Pause once more over the ordinary details. A flowing river can maintain a roughly stable level while water continually enters and leaves. A living cell can maintain internal concentrations through ongoing transport and reaction. Now imagine changing only one condition at a time: the material, the size, the surrounding medium, the duration of observation, or the character of the boundary. The visible event might remain recognizable, yet its pace and outcome could alter greatly. This is why a thermodynamic story becomes more accurate when it includes the setting instead of presenting the system as though it acted alone.

Language deserves its own small audit here. A stable reading may represent achieved equilibrium or sustained non-equilibrium, and the system history reveals which one it is. In everyday speech we often turn a process into a thing. We say that heat enters, pressure pushes, or equilibrium arrives. Such expressions are efficient, and there is no need to ban them. The useful habit is to ask what the noun stands for in the actual account: a measured relation, an interaction, a redistribution, a constraint, or a shorthand for several of these together.

Local and Global Equilibrium

Consider what an observer would actually notice. A small region of a large system may be close enough to equilibrium for temperature and pressure to be useful, while the larger system remains strongly non-uniform and evolving. None of this observation tells us, by itself, which metaphor must be used underneath it. The event gives us readings, timings, boundaries, and consequences. The deeper story is something we add in order to make those relations intelligible.

Thermodynamics and the relevant specialist science already organize the event. Continuum models often rely on local equilibrium approximations, while global gradients drive transport and large-scale change. These established descriptions are the common ground. They allow a teacher in one country, an engineer in another, and a researcher in a third field to speak about the same physical situation without depending on personal imagery.

Now introduce the Chavanian lens carefully. The Chavanian view sees nested architectures. One layer can be locally resolved while another carries persistent Delta across wider space and longer time. Instead of asking where an invisible content went, the lens asks how the system's condition, boundary, and available degrees of freedom shaped the response. The change of language is useful only if it produces greater discipline rather than a new ambiguity.

This way of speaking travels across disciplines with caution. Flames, atmospheres, stars, shock layers, biological tissues, and industrial flows all involve questions of scale. The recurring architecture may be similar even when the mechanisms, scales, and measurable variables are very different. A membrane is not a piston, a cell is not an engine, and an atmosphere is not a sealed laboratory vessel. Analogy helps only when it does not erase those distinctions.

A teacher can turn this into a conversation. Ask at what spatial and temporal resolution a state variable is being assigned and what larger gradients remain outside the local description. Ask which statement came from observation, which came from a tested model, and which came from interpretation. Students often discover that they understood the formal relation but had mixed several physical pictures together. Naming the layers gives them permission to separate those pictures without discarding the science.

The research boundary is equally important. Nested architecture is not a quantitative criterion for local equilibrium. Established transport and statistical methods must provide that criterion. U E D does not gain experimental status merely because it offers a coherent vocabulary. Structured Vacuum Energy and Alignment Threshold Time require independent definitions, connections to existing measurements, and possible tests. The present book keeps that obligation visible.

What, then, should remain after this section? Equilibrium is always a statement made about a system, a scale, a set of variables, and a permitted degree of precision. The Chavanian view invites another question after the standard explanation is complete: what architecture made this particular thermodynamic response possible?

Let us return to the event without technical haste. A small region of a large system may be close enough to equilibrium for temperature and pressure to be useful, while the larger system remains strongly non-uniform and evolving. An observer could record what happens, but a fuller explanation would also ask what had to be present before it could happen. There had to be a particular arrangement, a particular contact with the surroundings, and a particular range of possible responses. The event is therefore not only a change in a quantity. It is a change made possible by a physical situation.

This is where ordinary grammar can quietly mislead us. Equilibrium is always a statement made about a system, a scale, a set of variables, and a permitted degree of precision. A convenient sentence may give an abstract quantity the role of an actor, as though the quantity itself travelled, commanded, or chose. Scientists usually understand the shorthand, but new learners may build their intuition around the grammar. Rephrasing the sentence in terms of conditions, interactions, and measurable change often reveals what the original expression had compressed.

Closing Reflection

Equilibrium is not the elimination of all difference and not the freezing of all activity. It is an achieved condition defined with respect to particular variables, pathways, constraints, and scales. The Chavanian view emphasizes that structure can persist after one route of change has reached completion.

Equilibrium does not require a featureless world. It requires that the relevant thermodynamic Deltas no longer produce net change through the available pathways under the stated conditions.

Interlude II

The Same Cup in Three Rooms

Take one cup of water and imagine carrying it through three rooms. In the first room, the air is cooler than the water. In the second, the air is at the same temperature. In the third, the air is warmer. The cup has not changed identity, but its thermal relation to the surroundings has changed completely.

In the cooler room, the cup participates in a process that tends to reduce the thermal difference. In the equal-temperature room, there may be no net thermal exchange of the same kind, though microscopic interactions continue. In the warmer room, the direction reverses and the cup warms.

Temperature gives us the comparison. Boundary conditions give us the contact. Material properties and fluid motion influence the rate. The same object can therefore cool, remain approximately steady, or warm without acquiring three different thermodynamic personalities.

This scene helps separate state from relation. The cup has a temperature, but what happens next depends on the temperature of the surroundings and on the architecture joining them. A state coordinate matters within a relational field of conditions.

Pressure has the same broad lesson. A gas at one pressure behaves differently depending on the pressure beyond its boundary and whether the boundary can move. Concentration has the same broad lesson. A solution's next change depends on what lies across the membrane and whether the relevant species can pass.

The word equilibrium becomes clearer in the middle room. Equal temperature does not make the cup and room identical. It only tells us that the thermally relevant relation has reached a condition in which no net process of that kind is expected under the stated contact. Composition, mass, structure, and scale remain different.

Now imagine that the middle room is not stable. Sunlight enters through a window, an air conditioner cycles, people come and go, and the cup evaporates slowly. The phrase same temperature may be true only for a moment or only within measurement uncertainty. Equilibrium and steady state must always be attached to a timescale and a set of variables.

For a listener, the three rooms offer a compact memory. Temperature is not a command. It is a coordinate that allows comparison. The relation between conditions, together with the available pathway, gives thermodynamic direction.

For a scientist outside thermodynamics, the example provides a way to inspect disciplinary language. Whenever a variable is described as causing change, ask whether the actual explanation also requires a comparison, a gradient, a boundary, and a mechanism.

The Chavanian view uses Delta to hold that comparison in attention. It uses architecture to hold the pathway in attention.

It uses resolution to describe the event and consequence to describe the changed condition that follows.

None of these words replaces temperature, pressure, concentration, equilibrium, or the established laws. Their role is to prevent the physical narrative from collapsing into the image of a variable acting alone.

The cup in three rooms teaches a quiet principle: the meaning of a state becomes operational when the state is placed in relation to another condition and a physically available route.

Part III

The U E D Layer, Carefully Entered

→

Chapter 7

Persistence Within Continuity

What If Matter Is Persistence Within Continuity?

Objects Remain, but Separation Is Reconsidered

A molecule remains a molecule long enough to be studied. A crystal keeps an ordered form. A droplet holds an interface. A living cell preserves an identity while exchanging matter with its surroundings. The world contains persistent things. The question is whether their persistence requires them to be fundamentally isolated from the reality around them.

The question in this chapter is not whether established thermodynamics works. It is what we notice when we place the idea of architecture beside persistence, manifestation, continuity, and the proposed U E D foundation. The calculation and the measurement remain where science has placed them. Our task is to build a physical narrative that is clear enough for teaching, useful enough for interdisciplinary conversation, and modest enough to preserve the difference between interpretation and evidence.

We will move through familiar scenes, the established account, the Chavanian view, and the limits of that view. By the end, the phrase continuity need not erase structure, and structure need not imply absolute separation should feel less like a slogan and more like a disciplined question that can accompany technical study without competing with it.

The Object Is Not Denied

Consider what an observer would actually notice. A chair supports weight, an atom produces identifiable spectra, and a cell maintains a boundary. Treating these as manifestations does not make them imaginary or optional. None of this observation tells us, by itself, which metaphor must be used underneath it. The event gives us readings, timings, boundaries, and consequences. The deeper story is something we add in order to make those relations intelligible.

Thermodynamics and the relevant specialist science already organize the event. Physics and chemistry describe stable and metastable structures through fields, interactions, quantum states, bonding, collective behaviour, and environmental conditions. These established descriptions are the common ground. They allow a teacher in one country, an engineer in another, and a researcher in a third field to speak about the same physical situation without depending on personal imagery.

Now introduce the Chavanian lens carefully. The Chavanian view uses manifestation for an identifiable structured condition that persists within a wider reality. The term changes the account of separation, not the reality of the object. Instead of asking where an invisible content went, the lens asks how the system's condition, boundary, and available degrees of freedom shaped the response. The change of language is useful only if it produces greater discipline rather than a new ambiguity.

This way of speaking travels across disciplines with caution. Waves, vortices, defects in solids, flames, and biological organisms show different ways in which recognizable form can persist through continuing internal change. The recurring architecture may be similar even when the mechanisms, scales, and measurable variables are very different. A membrane is not a piston, a cell is not an engine, and an atmosphere is not a sealed laboratory vessel. Analogy helps only when it does not erase those distinctions.

A teacher can turn this into a conversation. Ask what makes an entity identifiable across time and which exchanges it can undergo without losing the identity relevant to the discussion. Ask which statement came from observation, which came from a tested model, and which came from interpretation. Students often discover that they understood the formal relation but had mixed several physical pictures together. Naming the layers gives them permission to separate those pictures without discarding the science.

The research boundary is equally important. Calling matter manifestation is not an established replacement for particle or field ontology. It is a philosophical and physical proposal within the Chavanian framework. U E D does not gain experimental status merely because it offers a coherent vocabulary. Structured Vacuum Energy and Alignment Threshold Time require independent definitions, connections to existing measurements, and possible tests. The present book keeps that obligation visible.

What, then, should remain after this section? Persistence makes an object real for science even if its deepest boundary is relational rather than absolutely separate. The Chavanian view invites another question after the standard explanation is complete: what architecture made this particular thermodynamic response possible?

Let us return to the event without technical haste. A chair supports weight, an atom produces identifiable spectra, and a cell maintains a boundary. Treating these as manifestations does not make them imaginary or optional. An observer could record what happens, but a fuller explanation would also ask what had to be present before it could happen. There had to be a particular arrangement, a particular contact with the surroundings, and a particular range of possible responses. The event is therefore not only a change in a quantity. It is a change made possible by a physical situation.

This is where ordinary grammar can quietly mislead us. Persistence makes an object real for science even if its deepest boundary is relational rather than absolutely separate. A convenient sentence may give an abstract quantity the role of an actor, as though the quantity itself travelled, commanded, or chose. Scientists usually understand the shorthand, but new learners may build their intuition around the grammar. Rephrasing the sentence in terms of conditions, interactions, and measurable change often reveals what the original expression had compressed.

Continuity Is Not Uniformity

A useful way to enter the topic is to slow the event down in imagination. An ocean is continuous across a wave, yet the wave is not identical to the calm water around it. A solid is continuous enough to transmit stress while containing grains, defects, surfaces, and internal structure. At each moment, the system has a condition and a set of possible next steps. Some steps are easy, some require a threshold, and some are excluded by the way the system is built or connected.

The established account tells us how to describe that condition responsibly. Continuum descriptions successfully represent many systems while allowing local fields, gradients, phases, interfaces, and discontinuities at chosen scales. It protects us from replacing measurement with intuition. Even in an equation-free conversation, the formal background matters because the words inherit their meaning from that disciplined practice.

From the Chavanian perspective, the event can be narrated as structured resolution. U E D extends continuity into a foundational proposal. Structured Vacuum Energy is imagined as the continuous physical setting within which persistent manifestations arise. Delta names the relevant difference. The architecture names the permitted and blocked routes. Resolution names the actual process. Consequence names the changed condition that remains after the process has occurred.

That sequence can illuminate another field. The distinction helps readers compare fields, media, order parameters, and emergent structures without assuming that all forms of continuity are the same. Yet the illumination is conceptual rather than a claim that one universal mechanism has already been demonstrated. The value lies in asking comparable questions: What persists? What differs? What pathway is available? What must the surroundings supply or receive?

In a classroom or research group, the vocabulary can be used as a checking device. Use a familiar continuous medium to show that continuity answers a question about connection, while uniformity answers a question about sameness. If the conversation begins to treat Delta as a force, architecture as a hidden machine, or resolution as a new measurable substance, the terms have drifted. Their purpose is to organize relations, not to populate nature with additional invisible objects.

The strongest caution is this. Ordinary continuum models do not provide evidence for Structured Vacuum Energy. The analogy explains the proposal but cannot verify it. Established thermodynamics can stand without the Chavanian interpretation. The interpretation must therefore earn its place through explanatory usefulness now and through mathematical and experimental work later. It cannot demand immunity from the evidence that shaped the field.

The chapter's central movement appears again. A continuous reality can be differentiated, structured, layered, and locally persistent. Change becomes easier to discuss when we keep both the initiating difference and the available pathway in view. The answer is not hidden in either one alone.

Imagine watching the scene with the ability to stop it at several moments. An ocean is continuous across a wave, yet the wave is not identical to the calm water around it. A solid is continuous enough to transmit stress while containing grains, defects, surfaces, and internal structure. At every pause, the system presents both a current condition and a limited range of next conditions. The range is shaped by material structure, contact, geometry, and surrounding constraints. Change is therefore neither a detached command nor a mysterious jump.

It occurs through what the present arrangement can physically support.

A small change in wording can expose this structure. A continuous reality can be differentiated, structured, layered, and locally persistent. Instead of asking what substance moved, ask which relation changed. Instead of asking what quantity wanted to increase or decrease, ask which processes were permitted under the stated conditions. The older wording may still be retained for convenience, but the second wording makes its hidden assumptions easier to inspect. That is especially helpful when familiar metaphors begin to conflict with one another.

Persistence Has a Timescale

Begin with the scene itself. A soap bubble persists for seconds, a weather system for days, a tree for decades, a mountain for millions of years, and some atomic structures far longer than human observation. It is easy to hurry past such an event because it is familiar. Yet familiarity often hides the number of conditions that must cooperate. The materials have properties. The surroundings have a history. A boundary permits some exchanges and suppresses others. The event looks simple only after all of those details have already been arranged.

Established science gives us the disciplined account. Stability, metastability, kinetics, environmental coupling, and scale determine how long a configuration remains identifiable. This account does not need a new vocabulary in order to remain valid. It tells us what to measure, which distinctions matter, and how the event can be compared with another event under specified conditions. Any deeper view must begin by respecting that achievement.

The Chavanian question begins one layer lower. The Chavanian view treats persistence as a successful accommodation of internal and external Deltas over a relevant timescale. Persistence is an ongoing condition, not an exemption from change. The word architecture directs attention away from an imagined travelling substance and toward the relations that make one response available while another remains blocked. This is an interpretation of the event, not a change to the event's measured description.

The same pattern becomes visible elsewhere. Materials ageing, protein folding, ecosystems, geological formations, and engineered structures all require attention to characteristic times. The comparison does not claim that the systems share an identical mechanism. It asks whether difference, constraint, pathway, and consequence provide a useful family resemblance across fields that often speak in separate dialects.

For teaching and early research, this distinction can be practical. Ask whether stable means forever, unchanged, self-maintaining, or merely long-lived compared with the observation. A learner can state the observation first, the accepted scientific relation second, and the proposed interpretation third. That order reduces the risk that a helpful picture will be remembered as though it were the definition itself.

There is also a limit that should remain audible. The general word persistence cannot replace domain-specific theories of stability and decay. A conceptual account becomes scientifically stronger only when its terms can be constrained by measurement and when it accepts the possibility of failure. Until then, clarity is valuable, but clarity is not confirmation.

The section therefore returns to the chapter question. Identity depends partly on the timescale at which the system is observed and the changes it can absorb without losing its defining structure. In the Chavanian view, a change is not understood only by naming what changed. We also ask which difference mattered, which route was open, and what new condition the response created.

Pause once more over the ordinary details. A soap bubble persists for seconds, a weather system for days, a tree for decades, a mountain for millions of years, and some atomic structures far longer than human observation. Now imagine changing only one condition at a time: the material, the size, the surrounding medium, the duration of observation, or the character of the boundary. The visible event might remain recognizable, yet its pace and outcome could alter greatly. This is why a thermodynamic story becomes more accurate when it includes the setting instead of presenting the system as though it acted alone.

Language deserves its own small audit here. Identity depends partly on the timescale at which the system is observed and the changes it can absorb without losing its defining structure. In everyday speech we often turn a process into a thing. We say that heat enters, pressure pushes, or equilibrium arrives. Such expressions are efficient, and there is no need to ban them. The useful habit is to ask what the noun stands for in the actual account: a measured relation, an interaction, a redistribution, a constraint, or a shorthand for several of these together.

The Transitional Boundary

Consider what an observer would actually notice. A droplet meets air through an interface with surface tension, composition changes, molecular exchange, and finite thickness. A cell membrane is a structured region, not a mathematical line. None of this observation tells us, by itself, which metaphor must be used underneath it. The event gives us readings, timings, boundaries, and consequences. The deeper story is something we add in order to make those relations intelligible.

Thermodynamics and the relevant specialist science already organize the event. Interfacial thermodynamics and materials science describe gradients, surface excess, adsorption, phase fronts, and boundary-specific properties. These established descriptions are the common ground. They allow a teacher in one country, an engineer in another, and a researcher in a third field to speak about the same physical situation without depending on personal imagery.

Now introduce the Chavanian lens carefully. The Chavanian view sees a boundary as a transition within continuity. Persistence can create a gradient without creating an unrelated universe on each side. Instead of asking where an invisible content went, the lens asks how the system's condition, boundary, and available degrees of freedom shaped the response. The change of language is useful only if it produces greater discipline rather than a new ambiguity.

This way of speaking travels across disciplines with caution. Grain boundaries, catalytic surfaces, biological membranes, contact zones, and atmospheric interfaces all make transitional regions important. The recurring architecture may be similar even when the mechanisms, scales, and measurable variables are very different. A membrane is not a piston, a cell is not an engine, and an atmosphere is not a sealed laboratory vessel. Analogy helps only when it does not erase those distinctions.

A teacher can turn this into a conversation. Ask what changes as one moves across the boundary and which properties belong specifically to the transitional region. Ask which statement came from observation, which came from a tested model, and which came from interpretation. Students often discover that they understood the formal relation but had mixed several physical pictures together. Naming the layers gives them permission to separate those pictures without discarding the science.

The research boundary is equally important. The move from measurable interfaces to foundational S V E continuity is a hypothesis, not a deduction from surface science. U E D does not gain experimental status merely because it offers a coherent vocabulary. Structured Vacuum Energy and Alignment Threshold Time require independent definitions, connections to existing measurements, and possible tests. The present book keeps that obligation visible.

What, then, should remain after this section? A boundary can preserve distinction while remaining physically connected to what lies around it. The Chavanian view invites another question after the standard explanation is complete: what architecture made this particular thermodynamic response possible?

Let us return to the event without technical haste. A droplet meets air through an interface with surface tension, composition changes, molecular exchange, and finite thickness. A cell membrane is a structured region, not a mathematical line. An observer could record what happens, but a fuller explanation would also ask what had to be present before it could happen. There had to be a particular arrangement, a particular contact with the surroundings, and a particular range of possible responses. The event is therefore not only a change in a quantity. It is a change made possible by a physical situation.

This is where ordinary grammar can quietly mislead us. A boundary can preserve distinction while remaining physically connected to what lies around it. A convenient sentence may give an abstract quantity the role of an actor, as though the quantity itself travelled, commanded, or chose. Scientists usually understand the shorthand, but new learners may build their intuition around the grammar. Rephrasing the sentence in terms of conditions, interactions, and measurable change often reveals what the original expression had compressed.

Why U E D Appears Here

A useful way to enter the topic is to slow the event down in imagination. Thermodynamics can be practised without deciding what ultimately fills the space between persistent structures. The measurements do not force one final ontology. At each moment, the system has a condition and a set of possible next steps. Some steps are easy, some require a threshold, and some are excluded by the way the system is built or connected.

The established account tells us how to describe that condition responsibly. Modern physics provides successful quantum field, particle, statistical, and continuum descriptions in their tested domains. Thermodynamics remains compatible with several microscopic interpretations. It protects us from replacing measurement with intuition. Even in an equation-free conversation, the formal background matters because the words inherit their meaning from that disciplined practice.

From the Chavanian perspective, the event can be narrated as structured resolution. U E D proposes Structured Vacuum Energy as a deeper continuous substrate and asks whether Delta and resolution can eventually be formalized within it. Delta names the relevant difference. The architecture names the permitted and blocked routes. Resolution names the actual process. Consequence names the changed condition that remains after the process has occurred.

That sequence can illuminate another field. The proposal touches questions in field theory, cosmology, gravitation, matter, and emergence, which is why its evidential obligations are much greater than those of a classroom metaphor. Yet the illumination is conceptual rather than a claim that one universal mechanism has already been demonstrated. The value lies in asking comparable questions: What persists? What differs? What pathway is available? What must the surroundings supply or receive?

In a classroom or research group, the vocabulary can be used as a checking device. Present U E D after the established explanation and ask students to identify what observation would remain unchanged and what new claim has been added. If the conversation begins to treat Delta as a force, architecture as a hidden machine, or resolution as a new measurable substance, the terms have drifted. Their purpose is to organize relations, not to populate nature with additional invisible objects.

The strongest caution is this. Structured Vacuum Energy is not the accepted vacuum of quantum field theory under another name, and it is not presently a complete predictive theory. Established thermodynamics can stand without the Chavanian interpretation. The interpretation must therefore earn its place through explanatory usefulness now and through mathematical and experimental work later. It cannot demand immunity from the evidence that shaped the field.

The chapter's central movement appears again. U E D is the proposed foundation beneath the Chavanian view, not a result already contained in thermodynamic measurements. Change becomes easier to discuss when we keep both the initiating difference and the available pathway in view. The answer is not hidden in either one alone.

Imagine watching the scene with the ability to stop it at several moments. Thermodynamics can be practised without deciding what ultimately fills the space between persistent structures. The measurements do not force one final ontology. At every pause, the system presents both a current condition and a limited range of next conditions. The range is shaped by material structure, contact, geometry, and surrounding constraints. Change is therefore neither a detached command nor a mysterious jump. It occurs through what the present arrangement can physically support.

A small change in wording can expose this structure. U E D is the proposed foundation beneath the Chavanian view, not a result already contained in thermodynamic measurements. Instead of asking what substance moved, ask which relation changed. Instead of asking what quantity wanted to increase or decrease, ask which processes were permitted under the stated conditions. The older wording may still be retained for convenience, but the second wording makes its hidden assumptions easier to inspect. That is especially helpful when familiar metaphors begin to conflict with one another.

Closing Reflection

The Chavanian view begins with persistent manifestations and asks whether their boundaries can be understood within continuity. U E D offers one proposed foundation for that continuity. The proposal is attractive because it gives difference and relation a physical setting. It remains provisional because attraction is not evidence and vocabulary is not yet mathematics.

Matter is not denied. The Chavanian proposal is that matter may be persistent structured manifestation within a continuous foundational reality rather than an isolated object placed in unrelated emptiness.

Chapter 8
The Available Door

What If Change Needs an Available Door?

A Difference Can Wait

Fuel and oxygen can remain together without immediate flame. A supercooled liquid can remain liquid below its usual freezing point. A compressed structure can persist until a crack begins. A difference may be present while the expected outcome waits. Something more than Delta is required.

The question in this chapter is not whether established thermodynamics works. It is what we notice when we place the idea of architecture beside thresholds, kinetics, access, competing routes, and system history. The calculation and the measurement remain where science has placed them. Our task is to build a physical narrative that is clear enough for teaching, useful enough for interdisciplinary conversation, and modest enough to preserve the difference between interpretation and evidence.

We will move through familiar scenes, the established account, the Chavanian view, and the limits of that view. By the end, the phrase a possible change becomes real only through a physically available pathway should feel less like a slogan and more like a disciplined question that can accompany technical study without competing with it.

Difference Without Immediate Change

A useful way to enter the topic is to slow the event down in imagination. A reaction can be thermodynamically favoured yet proceed imperceptibly slowly. A phase can remain metastable. A pressure difference can persist behind a strong wall. At each moment, the system has a condition and a set of possible next steps. Some steps are easy, some require a threshold, and some are excluded by the way the system is built or connected.

The established account tells us how to describe that condition responsibly. Kinetics, activation barriers, nucleation, transport limitations, mechanical strength, and boundary conditions determine whether and how quickly a process occurs. It protects us from replacing measurement with intuition. Even in an equation-free conversation, the formal background matters because the words inherit their meaning from that disciplined practice.

From the Chavanian perspective, the event can be narrated as structured resolution. The Chavanian view says Delta changes the field of possibility, while an available resolution geometry determines whether the possibility becomes operative. Delta names the relevant difference. The architecture names the permitted and blocked routes. Resolution names the actual process. Consequence names the changed condition that remains after the process has occurred.

That sequence can illuminate another field. Corrosion, combustion, protein folding, battery degradation, geological failure, and atmospheric nucleation all separate thermodynamic tendency from actual rate. Yet the illumination is conceptual rather than a claim that one universal mechanism has already been demonstrated. The value lies in asking comparable questions: What persists? What differs? What pathway is available? What must the surroundings supply or receive?

In a classroom or research group, the vocabulary can be used as a checking device. Ask two questions separately: is the change favoured under the constraints, and is there a pathway available on the timescale of interest? If the conversation begins to treat Delta as a force, architecture as a hidden machine, or resolution as a new measurable substance, the terms have drifted. Their purpose is to organize relations, not to populate nature with additional invisible objects.

The strongest caution is this. Resolution geometry is not a numerical rate law. It must defer to the established kinetics of the domain. Established thermodynamics can stand without the Chavanian interpretation. The interpretation must therefore earn its place through explanatory usefulness now and through mathematical and experimental work later. It cannot demand immunity from the evidence that shaped the field.

The chapter's central movement appears again. A thermodynamic direction does not guarantee an immediate event. Change becomes easier to discuss when we keep both the initiating difference and the available pathway in view. The answer is not hidden in either one alone.

Imagine watching the scene with the ability to stop it at several moments. A reaction can be thermodynamically favoured yet proceed imperceptibly slowly. A phase can remain metastable. A pressure difference can persist behind a strong wall. At every pause, the system presents both a current condition and a limited range of next conditions. The range is shaped by material structure, contact, geometry, and surrounding constraints. Change is therefore neither a detached command nor a mysterious jump. It occurs through what the present arrangement can physically support.

A small change in wording can expose this structure. A thermodynamic direction does not guarantee an immediate event. Instead of asking what substance moved, ask which relation changed. Instead of asking what quantity wanted to increase or decrease, ask which processes were permitted under the stated conditions. The older wording may still be retained for convenience, but the second wording makes its hidden assumptions easier to inspect. That is especially helpful when familiar metaphors begin to conflict with one another.

Thresholds and New Modes

Begin with the scene itself. A material can warm gradually, soften, transform phase, react, fracture, or ionize as conditions change. Different responses become accessible at different ranges. It is easy to hurry past such an event because it is familiar. Yet familiarity often hides the number of conditions that must cooperate. The materials have properties. The surroundings have a history. A boundary permits some exchanges and suppresses others. The event looks simple only after all of those details have already been arranged.

Established science gives us the disciplined account. Phase boundaries, activation energies, critical conditions, mechanical limits, and quantum transitions are defined through specific theories and measurements. This account does not need a new vocabulary in order to remain valid. It tells us what to measure, which distinctions matter, and how the event can be compared with another event under specified conditions. Any deeper view must begin by respecting that achievement.

The Chavanian question begins one layer lower. The Chavanian view speaks of layered accessibility. As conditions change, the architecture may open a mode of response that was previously unavailable. The word architecture directs attention away from an imagined travelling substance and toward the relations that make one response available while another remains blocked. This is an interpretation of the event, not a change to the event's measured description.

The same pattern becomes visible elsewhere. The idea connects materials processing, physiology, chemistry, geophysics, and plasma formation without making their thresholds equivalent. The comparison does not claim that the systems share an identical mechanism. It asks whether difference, constraint, pathway, and consequence provide a useful family resemblance across fields that often speak in separate dialects.

For teaching and early research, this distinction can be practical. Build a verbal map of possible responses and ask what condition opens each route and what evidence identifies the transition. A learner can state the observation first, the accepted scientific relation second, and the proposed interpretation third. That order reduces the risk that a helpful picture will be remembered as though it were the definition itself.

There is also a limit that should remain audible. Alignment Threshold Time is a proposed Chavanian concept. It must not be substituted for established threshold quantities or activation parameters. A conceptual account becomes scientifically stronger only when its terms can be constrained by measurement and when it accepts the possibility of failure. Until then, clarity is valuable, but clarity is not confirmation.

The section therefore returns to the chapter question. The outcome depends not only on how large a difference is but on what the system is physically able to do at that condition. In the Chavanian view, a change is not understood only by naming what changed. We also ask which difference mattered, which route was open, and what new condition the response created.

Pause once more over the ordinary details. A material can warm gradually, soften, transform phase, react, fracture, or ionize as conditions change. Different responses become accessible at different ranges. Now imagine changing only one condition at a time: the material, the size, the surrounding medium, the duration of observation, or the character of the boundary. The visible event might remain recognizable, yet its pace and outcome could alter greatly. This is why a thermodynamic story becomes more accurate when it includes the setting instead of presenting the system as though it acted alone.

Language deserves its own small audit here. The outcome depends not only on how large a difference is but on what the system is physically able to do at that condition. In everyday speech we often turn a process into a thing. We say that heat enters, pressure pushes, or equilibrium arrives. Such expressions are efficient, and there is no need to ban them. The useful habit is to ask what the noun stands for in the actual account: a measured relation, an interaction, a redistribution, a constraint, or a shorthand for several of these together.

Competing Pathways

Consider what an observer would actually notice. A hot object can cool by radiation, convection, conduction, evaporation, or several routes at once. A reacting molecule may encounter competing mechanisms and products. None of this observation tells us, by itself, which metaphor must be used underneath it. The event gives us readings, timings, boundaries, and consequences. The deeper story is something we add in order to make those relations intelligible.

Thermodynamics and the relevant specialist science already organize the event. Rates, geometry, material properties, coupling, and environmental conditions determine which processes dominate. These established descriptions are the common ground. They allow a teacher in one country, an engineer in another, and a researcher in a third field to speak about the same physical situation without depending on personal imagery.

Now introduce the Chavanian lens carefully. The Chavanian view describes the system as an architecture of simultaneous resolution possibilities. Consequence is distributed among the routes that are open and effective. Instead of asking where an invisible content went, the lens asks how the system's condition, boundary, and available degrees of freedom shaped the response. The change of language is useful only if it produces greater discipline rather than a new ambiguity.

This way of speaking travels across disciplines with caution. Parallel pathways appear in metabolism, atmospheric chemistry, catalysis, thermal management, and ecosystem processes. The recurring architecture may be similar even when the mechanisms, scales, and measurable variables are very different. A membrane is not a piston, a cell is not an engine, and an atmosphere is not a sealed laboratory vessel. Analogy helps only when it does not erase those distinctions.

A teacher can turn this into a conversation. Ask learners to list the routes before choosing the dominant one, then explain what experimental change would reorder them. Ask which statement came from observation, which came from a tested model, and which came from interpretation. Students often discover that they understood the formal relation but had mixed several physical pictures together. Naming the layers gives them permission to separate those pictures without discarding the science.

The research boundary is equally important. The architecture metaphor cannot calculate branching ratios or rates without a formal model. U E D does not gain experimental status merely because it offers a coherent vocabulary. Structured Vacuum Energy and Alignment Threshold Time require independent definitions, connections to existing measurements, and possible tests. The present book keeps that obligation visible.

What, then, should remain after this section? Nature need not select one pure response when several pathways can operate together. The Chavanian view invites another question after the standard explanation is complete: what architecture made this particular thermodynamic response possible?

Let us return to the event without technical haste. A hot object can cool by radiation, convection, conduction, evaporation, or several routes at once. A reacting molecule may encounter competing mechanisms and products. An observer could record what happens, but a fuller explanation would also ask what had to be present before it could happen. There had to be a particular arrangement, a particular contact with the surroundings, and a particular range of possible responses. The event is therefore not only a change in a quantity. It is a change made possible by a physical situation.

This is where ordinary grammar can quietly mislead us. Nature need not select one pure response when several pathways can operate together. A convenient sentence may give an abstract quantity the role of an actor, as though the quantity itself travelled, commanded, or chose. Scientists usually understand the shorthand, but new learners may build their intuition around the grammar. Rephrasing the sentence in terms of conditions, interactions, and measurable change often reveals what the original expression had compressed.

History and Hysteresis

A useful way to enter the topic is to slow the event down in imagination. Two systems can have the same present temperature and pressure yet respond differently because one was cooled, stressed, magnetized, dried, or transformed along a different path. At each moment, the system has a condition and a set of possible next steps. Some steps are easy, some require a threshold, and some are excluded by the way the system is built or connected.

The established account tells us how to describe that condition responsibly. Hysteresis, metastability, memory effects, structural defects, and path dependence appear across materials, phase transitions, chemistry, and complex systems. It protects us from replacing measurement with intuition. Even in an equation-free conversation, the formal background matters because the words inherit their meaning from that disciplined practice.

From the Chavanian perspective, the event can be narrated as structured resolution. The Chavanian view says changed geometry carries the consequence of earlier resolution. The present architecture contains a physical history. Delta names the relevant difference. The architecture names the permitted and blocked routes. Resolution names the actual process. Consequence names the changed condition that remains after the process has occurred.

That sequence can illuminate another field. Shape-memory materials, soils, polymers, proteins, batteries, and climate subsystems can all exhibit history-dependent response. Yet the illumination is conceptual rather than a claim that one universal mechanism has already been demonstrated. The value lies in asking comparable questions: What persists? What differs? What pathway is available? What must the surroundings supply or receive?

In a classroom or research group, the vocabulary can be used as a checking device. Ask whether the current state variables are sufficient for the question or whether hidden structural variables and prior pathways matter. If the conversation begins to treat Delta as a force, architecture as a hidden machine, or resolution as a new measurable substance, the terms have drifted. Their purpose is to organize relations, not to populate nature with additional invisible objects.

The strongest caution is this. History dependence is established in many domains. The Chavanian sequence is a way to narrate it, not a universal constitutive law. Established thermodynamics can stand without the Chavanian interpretation. The interpretation must therefore earn its place through explanatory usefulness now and through mathematical and experimental work later. It cannot demand immunity from the evidence that shaped the field.

The chapter's central movement appears again. What a system can do next may depend on what it has already done. Change becomes easier to discuss when we keep both the initiating difference and the available pathway in view. The answer is not hidden in either one alone.

Imagine watching the scene with the ability to stop it at several moments. Two systems can have the same present temperature and pressure yet respond differently because one was cooled, stressed, magnetized, dried, or transformed along a different path. At every pause, the system presents both a current condition and a limited range of next conditions. The range is shaped by material structure, contact, geometry, and surrounding constraints. Change is therefore neither a detached command nor a mysterious jump. It occurs through what the present arrangement can physically support.

A small change in wording can expose this structure. What a system can do next may depend on what it has already done. Instead of asking what substance moved, ask which relation changed. Instead of asking what quantity wanted to increase or decrease, ask which processes were permitted under the stated conditions. The older wording may still be retained for convenience, but the second wording makes its hidden assumptions easier to inspect. That is especially helpful when familiar metaphors begin to conflict with one another.

Opening and Closing Doors

Begin with the scene itself. A catalyst opens a faster chemical route, insulation suppresses thermal routes, a valve changes flow geometry, and an enzyme makes a biological reaction accessible under cellular conditions. It is easy to hurry past such an event because it is familiar. Yet familiarity often hides the number of conditions that must cooperate. The materials have properties. The surroundings have a history. A boundary permits some exchanges and suppresses others. The event looks simple only after all of those details have already been arranged.

Established science gives us the disciplined account. Catalysts alter kinetics without changing the equilibrium condition; valves and insulation modify transport and boundary conditions; enzymes provide specific molecular mechanisms. This account does not need a new vocabulary in order to remain valid. It tells us what to measure, which distinctions matter, and how the event can be compared with another event under specified conditions. Any deeper view must begin by respecting that achievement.

The Chavanian question begins one layer lower. The Chavanian view groups these events as changes in resolution accessibility while preserving their distinct mechanisms. The word architecture directs attention away from an imagined travelling substance and toward the relations that make one response available while another remains blocked. This is an interpretation of the event, not a change to the event's measured description.

The same pattern becomes visible elsewhere. Engineering and biology often succeed by controlling pathways rather than by changing the final thermodynamic criterion. The comparison does not claim that the systems share an identical mechanism. It asks whether difference, constraint, pathway, and consequence provide a useful family resemblance across fields that often speak in separate dialects.

For teaching and early research, this distinction can be practical. Ask whether an intervention changes the initial state, the permitted route, the rate, the final equilibrium, or several of these. A learner can state the observation first, the accepted scientific relation second, and the proposed interpretation third. That order reduces the risk that a helpful picture will be remembered as though it were the definition itself.

There is also a limit that should remain audible. Calling a catalyst a door is metaphorical. Mechanistic chemistry remains necessary. A conceptual account becomes scientifically stronger only when its terms can be constrained by measurement and when it accepts the possibility of failure. Until then, clarity is valuable, but clarity is not confirmation.

The section therefore returns to the chapter question. Control often works by reorganizing the route through which an existing difference can become consequence. In the Chavanian view, a change is not understood only by naming what changed. We also ask which difference mattered, which route was open, and what new condition the response created.

Pause once more over the ordinary details. A catalyst opens a faster chemical route, insulation suppresses thermal routes, a valve changes flow geometry, and an enzyme makes a biological reaction accessible under cellular conditions. Now imagine changing only one condition at a time: the material, the size, the surrounding medium, the duration of observation, or the character of the boundary. The visible event might remain recognizable, yet its pace and outcome could alter greatly. This is why a thermodynamic story becomes more accurate when it includes the setting instead of presenting the system as though it acted alone.

Language deserves its own small audit here. Control often works by reorganizing the route through which an existing difference can become consequence. In everyday speech we often turn a process into a thing. We say that heat enters, pressure pushes, or equilibrium arrives. Such expressions are efficient, and there is no need to ban them. The useful habit is to ask what the noun stands for in the actual account: a measured relation, an interaction, a redistribution, a constraint, or a shorthand for several of these together.

Closing Reflection

The existence of Delta does not guarantee that a change will occur on the timescale we care about. Systems have barriers, thresholds, competing routes, and histories. The Chavanian idea of an available door brings these conditions into one story, while the technical sciences remain responsible for identifying and calculating the actual mechanisms.

Difference makes change relevant. Accessibility makes change possible. Kinetics and mechanism determine how the possibility becomes an event.

Chapter 9
Entropy as Record

What If Entropy Records Distributed Consequence?

Disorder Is Memorable, but It Is Not the Whole Account

Entropy is often introduced through disorder. The image is memorable: a tidy room becomes untidy, a drop of ink spreads through water, and a broken cup does not rebuild itself. But thermodynamic entropy is more disciplined than the ordinary meaning of mess. The challenge is to preserve the science while improving the story.

The question in this chapter is not whether established thermodynamics works. It is what we notice when we place the idea of architecture beside entropy, multiplicity, irreversibility, information, and the Second Law. The calculation and the measurement remain where science has placed them. Our task is to build a physical narrative that is clear enough for teaching, useful enough for interdisciplinary conversation, and modest enough to preserve the difference between interpretation and evidence.

We will move through familiar scenes, the established account, the Chavanian view, and the limits of that view. By the end, the phrase entropy remains an established state quantity while the Chavanian view asks what physical distribution it records should feel less like a slogan and more like a disciplined question that can accompany technical study without competing with it.

Why Disorder Helps and Misleads

Begin with the scene itself. Mixing, diffusion, and spreading look less ordered to human eyes, so disorder seems to provide an intuitive arrow. Yet a crystal can have defects, a gas can be spatially patterned, and a living system can be visually organized while producing entropy. It is easy to hurry past such an event because it is familiar. Yet familiarity often hides the number of conditions that must cooperate. The materials have properties. The surroundings have a history. A boundary permits some exchanges and suppresses others. The event looks simple only after all of those details have already been arranged.

Established science gives us the disciplined account. Entropy is a thermodynamic state quantity with statistical interpretations connected to the multiplicity and probability of microscopic configurations under defined constraints. This account does not need a new vocabulary in order to remain valid. It tells us what to measure, which distinctions matter, and how the event can be compared with another event under specified conditions. Any deeper view must begin by respecting that achievement.

The Chavanian question begins one layer lower. The Chavanian view asks whether entropy can be narrated as a record of how consequence becomes distributed across accessible microscopic and macroscopic relations. The word architecture directs attention away from an imagined travelling substance and toward the relations that make one response available while another remains blocked. This is an interpretation of the event, not a change to the event's measured description.

The same pattern becomes visible elsewhere. The distinction matters in chemistry, information science, biology, cosmology, and public explanations where disorder is easily stretched beyond its technical role. The comparison does not claim that the systems share an identical mechanism. It asks whether difference, constraint, pathway, and consequence provide a useful family resemblance across fields that often speak in separate dialects.

For teaching and early research, this distinction can be practical. Use examples in which visual tidiness and thermodynamic entropy do not track one another, then return to the formal definition used in the course. A learner can state the observation first, the accepted scientific relation second, and the proposed interpretation third. That order reduces the risk that a helpful picture will be remembered as though it were the definition itself.

There is also a limit that should remain audible. Distributed consequence is an interpretation, not a new entropy equation or a replacement for statistical mechanics. A conceptual account becomes scientifically stronger only when its terms can be constrained by measurement and when it accepts the possibility of failure. Until then, clarity is valuable, but clarity is not confirmation.

The section therefore returns to the chapter question. Disorder may illustrate some cases, but entropy cannot be reduced to what looks messy to an observer. In the Chavanian view, a change is not understood only by naming what changed. We also ask which difference mattered, which route was open, and what new condition the response created.

Pause once more over the ordinary details. Mixing, diffusion, and spreading look less ordered to human eyes, so disorder seems to provide an intuitive arrow. Yet a crystal can have defects, a gas can be spatially patterned, and a living system can be visually organized while producing entropy. Now imagine changing only one condition at a time: the material, the size, the surrounding medium, the duration of observation, or the character of the boundary. The visible event might remain recognizable, yet its pace and outcome could alter greatly. This is why a thermodynamic story becomes more accurate when it includes the setting instead of presenting the system as though it acted alone.

Language deserves its own small audit here. Disorder may illustrate some cases, but entropy cannot be reduced to what looks messy to an observer. In everyday speech we often turn a process into a thing. We say that heat enters, pressure pushes, or equilibrium arrives. Such expressions are efficient, and there is no need to ban them. The useful habit is to ask what the noun stands for in the actual account: a measured relation, an interaction, a redistribution, a constraint, or a shorthand for several of these together.

Multiplicity Without Fundamental Chaos

Consider what an observer would actually notice. A macroscopic gas state can correspond to an immense number of microscopic configurations. We use statistical description because following every constituent is impossible and often unnecessary. None of this observation tells us, by itself, which metaphor must be used underneath it. The event gives us readings, timings, boundaries, and consequences. The deeper story is something we add in order to make those relations intelligible.

Thermodynamics and the relevant specialist science already organize the event. Statistical mechanics relates thermodynamic behaviour to ensembles, probabilities, state counting, indistinguishability, and the constraints defining accessible microstates. These established descriptions are the common ground. They allow a teacher in one country, an engineer in another, and a researcher in a third field to speak about the same physical situation without depending on personal imagery.

Now introduce the Chavanian lens carefully. The Chavanian view accepts the statistical structure while leaving open whether randomness is fundamental or reflects distributed deterministic relations within a deeper continuous reality. Instead of asking where an invisible content went, the lens asks how the system's condition, boundary, and available degrees of freedom shaped the response. The change of language is useful only if it produces greater discipline rather than a new ambiguity.

This way of speaking travels across disciplines with caution. The question resembles interpretive debates elsewhere in physics, where the success of a probability description does not by itself settle the deepest ontology. The recurring architecture may be similar even when the mechanisms, scales, and measurable variables are very different. A membrane is not a piston, a cell is not an engine, and an atmosphere is not a sealed laboratory vessel. Analogy helps only when it does not erase those distinctions.

A teacher can turn this into a conversation. Separate the measured success of statistical mechanics from the philosophical claim that microscopic reality must be fundamentally random in one particular sense. Ask which statement came from observation, which came from a tested model, and which came from interpretation. Students often discover that they understood the formal relation but had mixed several physical pictures together. Naming the layers gives them permission to separate those pictures without discarding the science.

The research boundary is equally important. U E D has not yet recovered the quantitative achievements of statistical mechanics from deeper principles. It cannot claim priority over them. U E D does not gain experimental status merely because it offers a coherent vocabulary. Structured Vacuum Energy and Alignment Threshold Time require independent definitions, connections to existing measurements, and possible tests. The present book keeps that obligation visible.

What, then, should remain after this section? A statistical description can be indispensable even while foundational questions about randomness remain open. The Chavanian view invites another question after the standard explanation is complete: what architecture made this particular thermodynamic response possible?

Let us return to the event without technical haste. A macroscopic gas state can correspond to an immense number of microscopic configurations. We use statistical description because following every constituent is impossible and often unnecessary. An observer could record what happens, but a fuller explanation would also ask what had to be present before it could happen. There had to be a particular arrangement, a particular contact with the surroundings, and a particular range of possible responses. The event is therefore not only a change in a quantity. It is a change made possible by a physical situation.

This is where ordinary grammar can quietly mislead us. A statistical description can be indispensable even while foundational questions about randomness remain open. A convenient sentence may give an abstract quantity the role of an actor, as though the quantity itself travelled, commanded, or chose. Scientists usually understand the shorthand, but new learners may build their intuition around the grammar. Rephrasing the sentence in terms of conditions, interactions, and measurable change often reveals what the original expression had compressed.

Irreversibility and Wider Distribution

A useful way to enter the topic is to slow the event down in imagination. Friction turns directed motion into distributed internal change. Mixing removes an easily recoverable separation. Heat passing through a finite temperature difference leaves less opportunity for directed use. At each moment, the system has a condition and a set of possible next steps. Some steps are easy, some require a threshold, and some are excluded by the way the system is built or connected.

The established account tells us how to describe that condition responsibly. Entropy production quantifies irreversibility in established thermodynamics and non-equilibrium theory. Energy accounting remains conserved while useful constraints and gradients change. It protects us from replacing measurement with intuition. Even in an equation-free conversation, the formal background matters because the words inherit their meaning from that disciplined practice.

From the Chavanian perspective, the event can be narrated as structured resolution. The Chavanian view says consequence becomes more broadly distributed through the available architecture, making exact reversal physically inaccessible without additional changes elsewhere. Delta names the relevant difference. The architecture names the permitted and blocked routes. Resolution names the actual process. Consequence names the changed condition that remains after the process has occurred.

That sequence can illuminate another field. Turbulence, electrical resistance, viscous flow, chemical reaction, diffusion, and biological metabolism all produce entropy through distinct mechanisms. Yet the illumination is conceptual rather than a claim that one universal mechanism has already been demonstrated. The value lies in asking comparable questions: What persists? What differs? What pathway is available? What must the surroundings supply or receive?

In a classroom or research group, the vocabulary can be used as a checking device. Ask where directed constraints were lost, what new microscopic distributions appeared, and what would have to be controlled to reverse the process. If the conversation begins to treat Delta as a force, architecture as a hidden machine, or resolution as a new measurable substance, the terms have drifted. Their purpose is to organize relations, not to populate nature with additional invisible objects.

The strongest caution is this. Inaccessibility should not be described as absolute metaphysical impossibility without specifying scale, fluctuations, and conditions. Established thermodynamics can stand without the Chavanian interpretation. The interpretation must therefore earn its place through explanatory usefulness now and through mathematical and experimental work later. It cannot demand immunity from the evidence that shaped the field.

The chapter's central movement appears again. Irreversibility concerns the changed accessibility of pathways, not the disappearance of energy from the ledger. Change becomes easier to discuss when we keep both the initiating difference and the available pathway in view. The answer is not hidden in either one alone.

Imagine watching the scene with the ability to stop it at several moments. Friction turns directed motion into distributed internal change. Mixing removes an easily recoverable separation. Heat passing through a finite temperature difference leaves less opportunity for directed use. At every pause, the system presents both a current condition and a limited range of next conditions. The range is shaped by material structure, contact, geometry, and surrounding constraints. Change is therefore neither a detached command nor a mysterious jump. It occurs through what the present arrangement can physically support.

A small change in wording can expose this structure. Irreversibility concerns the changed accessibility of pathways, not the disappearance of energy from the ledger. Instead of asking what substance moved, ask which relation changed. Instead of asking what quantity wanted to increase or decrease, ask which processes were permitted under the stated conditions. The older wording may still be retained for convenience, but the second wording makes its hidden assumptions easier to inspect. That is especially helpful when familiar metaphors begin to conflict with one another.

Entropy and Information

Begin with the scene itself. Information language enters thermodynamics through measurement, state description, uncertainty, computation, and the physical cost of manipulating records. It can clarify and confuse in equal measure. It is easy to hurry past such an event because it is familiar. Yet familiarity often hides the number of conditions that must cooperate. The materials have properties. The surroundings have a history. A boundary permits some exchanges and suppresses others. The event looks simple only after all of those details have already been arranged.

Established science gives us the disciplined account. Statistical entropy and information measures have precise mathematical relations in defined contexts, but their physical interpretations depend on the system and constraints. This account does not need a new vocabulary in order to remain valid. It tells us what to measure, which distinctions matter, and how the event can be compared with another event under specified conditions. Any deeper view must begin by respecting that achievement.

The Chavanian question begins one layer lower. The Chavanian view insists that observer description and physical accessibility must not be silently merged. A label invented by an observer does not create a new physical microstate. The word architecture directs attention away from an imagined travelling substance and toward the relations that make one response available while another remains blocked. This is an interpretation of the event, not a change to the event's measured description.

The same pattern becomes visible elsewhere. The issue appears in computing, genetics, neuroscience, black-hole discussions, and complex systems, where the word information carries several meanings. The comparison does not claim that the systems share an identical mechanism. It asks whether difference, constraint, pathway, and consequence provide a useful family resemblance across fields that often speak in separate dialects.

For teaching and early research, this distinction can be practical. Ask whether information refers to a physical record, a probability distribution, an observer's knowledge, a communication measure, or a structural distinction in the system. A learner can state the observation first, the accepted scientific relation second, and the proposed interpretation third. That order reduces the risk that a helpful picture will be remembered as though it were the definition itself.

There is also a limit that should remain audible. The Chavanian framework does not settle the physics of information by vocabulary. Each use requires its own operational definition. A conceptual account becomes scientifically stronger only when its terms can be constrained by measurement and when it accepts the possibility of failure. Until then, clarity is valuable, but clarity is not confirmation.

The section therefore returns to the chapter question. Physical distinctions and descriptions of those distinctions are related, but they are not automatically the same thing. In the Chavanian view, a change is not understood only by naming what changed. We also ask which difference mattered, which route was open, and what new condition the response created.

Pause once more over the ordinary details. Information language enters thermodynamics through measurement, state description, uncertainty, computation, and the physical cost of manipulating records. It can clarify and confuse in equal measure. Now imagine changing only one condition at a time: the material, the size, the surrounding medium, the duration of observation, or the character of the boundary. The visible event might remain recognizable, yet its pace and outcome could alter greatly. This is why a thermodynamic story becomes more accurate when it includes the setting instead of presenting the system as though it acted alone.

Language deserves its own small audit here. Physical distinctions and descriptions of those distinctions are related, but they are not automatically the same thing. In everyday speech we often turn a process into a thing. We say that heat enters, pressure pushes, or equilibrium arrives. Such expressions are efficient, and there is no need to ban them. The useful habit is to ask what the noun stands for in the actual account: a measured relation, an interaction, a redistribution, a constraint, or a shorthand for several of these together.

The Second Law Without an Invisible Enforcer

Consider what an observer would actually notice. Heat does not spontaneously concentrate from a cooler body into a warmer one in an isolated everyday process, and a cyclic machine cannot convert all thermal input into work without other consequence. None of this observation tells us, by itself, which metaphor must be used underneath it. The event gives us readings, timings, boundaries, and consequences. The deeper story is something we add in order to make those relations intelligible.

Thermodynamics and the relevant specialist science already organize the event. The Second Law gives precise directional restrictions and entropy criteria supported by extensive observation and engineering practice. These established descriptions are the common ground. They allow a teacher in one country, an engineer in another, and a researcher in a third field to speak about the same physical situation without depending on personal imagery.

Now introduce the Chavanian lens carefully. The Chavanian view interprets the direction as the statistical and structural tendency of unresolved Deltas to use available pathways and redistribute consequence. Instead of asking where an invisible content went, the lens asks how the system's condition, boundary, and available degrees of freedom shaped the response. The change of language is useful only if it produces greater discipline rather than a new ambiguity.

This way of speaking travels across disciplines with caution. Engines, chemical reactions, diffusion, climate processes, and living metabolism all operate within the Second Law rather than around it. The recurring architecture may be similar even when the mechanisms, scales, and measurable variables are very different. A membrane is not a piston, a cell is not an engine, and an atmosphere is not a sealed laboratory vessel. Analogy helps only when it does not erase those distinctions.

A teacher can turn this into a conversation. State the law in its accepted form first, then ask what physical picture makes the directional restriction intuitive without turning entropy into an agent. Ask which statement came from observation, which came from a tested model, and which came from interpretation. Students often discover that they understood the formal relation but had mixed several physical pictures together. Naming the layers gives them permission to separate those pictures without discarding the science.

The research boundary is equally important. An Equilibrium Principle or resolution narrative has scientific value only if it recovers the full quantitative Second Law and does not create forbidden exceptions. U E D does not gain experimental status merely because it offers a coherent vocabulary. Structured Vacuum Energy and Alignment Threshold Time require independent definitions, connections to existing measurements, and possible tests. The present book keeps that obligation visible.

What, then, should remain after this section? The Second Law needs no police officer inside nature. It describes the reliable direction of accessible change under defined conditions. The Chavanian view invites another question after the standard explanation is complete: what architecture made this particular thermodynamic response possible?

Let us return to the event without technical haste. Heat does not spontaneously concentrate from a cooler body into a warmer one in an isolated everyday process, and a cyclic machine cannot convert all thermal input into work without other consequence. An observer could record what happens, but a fuller explanation would also ask what had to be present before it could happen. There had to be a particular arrangement, a particular contact with the surroundings, and a particular range of possible responses. The event is therefore not only a change in a quantity. It is a change made possible by a physical situation.

This is where ordinary grammar can quietly mislead us. The Second Law needs no police officer inside nature. It describes the reliable direction of accessible change under defined conditions. A convenient sentence may give an abstract quantity the role of an actor, as though the quantity itself travelled, commanded, or chose. Scientists usually understand the shorthand, but new learners may build their intuition around the grammar. Rephrasing the sentence in terms of conditions, interactions, and measurable change often reveals what the original expression had compressed.

Closing Reflection

Entropy is too important to be reduced either to a slogan about disorder or to a new Chavanian label. The established quantity remains where thermodynamics and statistical mechanics have placed it. The Chavanian view contributes a question: might entropy record how thermodynamic consequence becomes distributed across accessible configurations and pathways? That question must eventually meet mathematics.

Entropy remains an established thermodynamic state quantity. The Chavanian interpretation treats its increase as a record of increasingly distributed consequence, not as the action of a substance or agent called disorder.

Interlude III

The Refrigerator That Did Not Break the Second Law

A child opens a refrigerator and feels cold air. The shelves are cool, the food is preserved, and the kitchen remains comfortable. It is tempting to look only inside the compartment and conclude that the machine has made heat move in the wrong direction.

Walk behind the appliance and the story changes. The coils are warm. The compressor consumes electrical work. The room receives thermal consequence from the refrigerated space together with the consequences of the work supplied to the machine.

The cold interior is real, but it is local. The machine maintains that local condition by participating in a wider cycle. The Second Law is not broken; the correct system was initially too narrow.

This is one of the most useful thermodynamic habits a teacher can cultivate. When an event appears to reverse an expected direction, widen the boundary. Ask what work is supplied, what reservoirs are involved, what matter crosses, and where the wider consequence appears.

The habit is equally valuable in biology. A cell maintains internal composition and organized structure, but it does so through metabolism, transport, exchange, and dissipation. The local persistence is not a sealed exception to thermodynamic direction.

It is valuable in climate discussions. A region can cool while the planet gains thermal consequence, or a device can improve local efficiency while shifting effects to another part of its life cycle. The chosen boundary determines which part of the event is visible.

The Chavanian view describes the refrigerator as an engineered resolution geometry. The machine does not remove the thermal Delta once and for all. It repeatedly manages pathways, using work to sustain a difference that would otherwise diminish.

The phrase engineered resolution is not an efficiency formula and not a new refrigeration cycle. It is a conceptual description of what the established cycle achieves. Engineers still need the correct refrigerant properties, pressures, temperatures, components, controls, safety standards, and calculations.

Universal Energy Dynamics is even farther beneath the appliance. It asks whether the entire connected process occurs within a continuous physical foundation. The humming compressor does not prove that proposal. It merely supplies a familiar event in which continuity, relation, and boundary can be discussed.

The refrigerator also corrects a common image of equilibrium. Nature does not demand that every local difference disappear immediately. Differences can be maintained when an open architecture continually creates the conditions for their persistence and produces consequence in a wider domain.

There is no free lunch hidden in the phrase. Maintenance has a cost. Real machines dissipate. Components age. Electrical generation has upstream conditions. The wider the boundary, the more complete the story becomes.

The refrigerator therefore carries us from the entropy discussion toward machines and research. It shows how a familiar device can support several layers of explanation at once: accepted thermodynamics, a Chavanian organizational view, and a provisional foundational proposal. Scientific honesty depends on keeping the layers visible even when the story flows smoothly.

Part IV

Across Fields and Toward Research

Chapter 10
Machines Organize Pathways

What If Machines Organize the Way Differences Resolve?

The Engine Does Not Defeat the Gradient

A machine can make a thermal difference useful. A heat engine turns part of a process into directed work. A refrigerator spends work to maintain a cooler region. The machine appears to negotiate with thermodynamic direction, but it never leaves the larger accounting behind.

The question in this chapter is not whether established thermodynamics works. It is what we notice when we place the idea of architecture beside engines, refrigerators, cycles, dissipation, and designed pathways. The calculation and the measurement remain where science has placed them. Our task is to build a physical narrative that is clear enough for teaching, useful enough for interdisciplinary conversation, and modest enough to preserve the difference between interpretation and evidence.

We will move through familiar scenes, the established account, the Chavanian view, and the limits of that view. By the end, the phrase a machine channels consequence without escaping thermodynamic limits should feel less like a slogan and more like a disciplined question that can accompany technical study without competing with it.

The Heat Engine as Organized Path

Consider what an observer would actually notice. A heat engine operates between regions at different temperatures and uses a cycle to produce directed mechanical consequence while rejecting thermal consequence elsewhere. None of this observation tells us, by itself, which metaphor must be used underneath it. The event gives us readings, timings, boundaries, and consequences. The deeper story is something we add in order to make those relations intelligible.

Thermodynamics and the relevant specialist science already organize the event. Established thermodynamics defines efficiencies, cycle relations, reservoir interactions, work, heat, and the limits imposed by the Second Law. These established descriptions are the common ground. They allow a teacher in one country, an engineer in another, and a researcher in a third field to speak about the same physical situation without depending on personal imagery.

Now introduce the Chavanian lens carefully. The Chavanian view describes the engine as a designed resolution geometry that channels part of a thermal Delta into a directed boundary process. Instead of asking where an invisible content went, the lens asks how the system's condition, boundary, and available degrees of freedom shaped the response. The change of language is useful only if it produces greater discipline rather than a new ambiguity.

This way of speaking travels across disciplines with caution. Steam plants, internal-combustion engines, turbines, thermoelectric devices, and biological molecular machines organize different mechanisms within thermodynamic limits. The recurring architecture may be similar even when the mechanisms, scales, and measurable variables are very different. A membrane is not a piston, a cell is not an engine, and an atmosphere is not a sealed laboratory vessel. Analogy helps only when it does not erase those distinctions.

A teacher can turn this into a conversation. Ask where the initial difference comes from, which route creates useful work, and where the remaining consequence goes. Ask which statement came from observation, which came from a tested model, and which came from interpretation. Students often discover that they understood the formal relation but had mixed several physical pictures together. Naming the layers gives them permission to separate those pictures without discarding the science.

The research boundary is equally important. The word channel does not imply that all thermal energy could become work with better ingenuity. The Second Law remains decisive. U E D does not gain experimental status merely because it offers a coherent vocabulary. Structured Vacuum Energy and Alignment Threshold Time require independent definitions, connections to existing measurements, and possible tests. The present book keeps that obligation visible.

What, then, should remain after this section? A machine gains usefulness by organizing the path, not by abolishing the need for a gradient and a wider consequence. The Chavanian view invites another question after the standard explanation is complete: what architecture made this particular thermodynamic response possible?

Let us return to the event without technical haste. A heat engine operates between regions at different temperatures and uses a cycle to produce directed mechanical consequence while rejecting thermal consequence elsewhere. An observer could record what happens, but a fuller explanation would also ask what had to be present before it could happen. There had to be a particular arrangement, a particular contact with the surroundings, and a particular range of possible responses. The event is therefore not only a change in a quantity. It is a change made possible by a physical situation.

This is where ordinary grammar can quietly mislead us. A machine gains usefulness by organizing the path, not by abolishing the need for a gradient and a wider consequence. A convenient sentence may give an abstract quantity the role of an actor, as though the quantity itself travelled, commanded, or chose. Scientists usually understand the shorthand, but new learners may build their intuition around the grammar. Rephrasing the sentence in terms of conditions, interactions, and measurable change often reveals what the original expression had compressed.

The Refrigerator and the Wider Room

A useful way to enter the topic is to slow the event down in imagination. Inside a refrigerator, food remains cooler than the kitchen. Behind the appliance, a warm coil releases thermal consequence to the room while electrical work supports the cycle. At each moment, the system has a condition and a set of possible next steps. Some steps are easy, some require a threshold, and some are excluded by the way the system is built or connected.

The established account tells us how to describe that condition responsibly. Refrigeration transfers heat from a colder region to a warmer environment by consuming work and rejecting a larger thermal amount to the surroundings. It protects us from replacing measurement with intuition. Even in an equation-free conversation, the formal background matters because the words inherit their meaning from that disciplined practice.

From the Chavanian perspective, the event can be narrated as structured resolution. The Chavanian view sees a locally maintained Delta embedded in a wider architecture that continuously creates consequence elsewhere. Delta names the relevant difference. The architecture names the permitted and blocked routes. Resolution names the actual process. Consequence names the changed condition that remains after the process has occurred.

That sequence can illuminate another field. Heat pumps, air conditioning, cryogenic cooling, data-centre cooling, and biological thermoregulation all require a widened boundary for honest interpretation. Yet the illumination is conceptual rather than a claim that one universal mechanism has already been demonstrated. The value lies in asking comparable questions: What persists? What differs? What pathway is available? What must the surroundings supply or receive?

In a classroom or research group, the vocabulary can be used as a checking device. Follow the process around the entire cycle and include the room or external reservoir rather than looking only at the cold compartment. If the conversation begins to treat Delta as a force, architecture as a hidden machine, or resolution as a new measurable substance, the terms have drifted. Their purpose is to organize relations, not to populate nature with additional invisible objects.

The strongest caution is this. A refrigerator is not a model of life, and local order is not automatically a violation of entropy increase in the wider system. Established thermodynamics can stand without the Chavanian interpretation. The interpretation must therefore earn its place through explanatory usefulness now and through mathematical and experimental work later. It cannot demand immunity from the evidence that shaped the field.

The chapter's central movement appears again. Local thermal difference can be maintained when work and wider environmental consequence are included. Change becomes easier to discuss when we keep both the initiating difference and the available pathway in view. The answer is not hidden in either one alone.

Imagine watching the scene with the ability to stop it at several moments. Inside a refrigerator, food remains cooler than the kitchen. Behind the appliance, a warm coil releases thermal consequence to the room while electrical work supports the cycle. At every pause, the system presents both a current condition and a limited range of next conditions. The range is shaped by material structure, contact, geometry, and surrounding constraints. Change is therefore neither a detached command nor a mysterious jump. It occurs through what the present arrangement can physically support.

A small change in wording can expose this structure. Local thermal difference can be maintained when work and wider environmental consequence are included. Instead of asking what substance moved, ask which relation changed. Instead of asking what quantity wanted to increase or decrease, ask which processes were permitted under the stated conditions. The older wording may still be retained for convenience, but the second wording makes its hidden assumptions easier to inspect. That is especially helpful when familiar metaphors begin to conflict with one another.

Dissipation and Lost Direction

Begin with the scene itself. Bearings heat, wires resist, fluids mix, shocks form, and turbulence spreads organized motion across many smaller scales. The energy balance can still close while useful performance declines. It is easy to hurry past such an event because it is familiar. Yet familiarity often hides the number of conditions that must cooperate. The materials have properties. The surroundings have a history. A boundary permits some exchanges and suppresses others. The event looks simple only after all of those details have already been arranged.

Established science gives us the disciplined account. Friction, viscosity, resistance, finite gradients, mixing, and other irreversible processes produce entropy and reduce the maximum work available from a given condition. This account does not need a new vocabulary in order to remain valid. It tells us what to measure, which distinctions matter, and how the event can be compared with another event under specified conditions. Any deeper view must begin by respecting that achievement.

The Chavanian question begins one layer lower. The Chavanian view describes dissipation as consequence becoming distributed into routes that are less available for controlled return to the original directed task. The word architecture directs attention away from an imagined travelling substance and toward the relations that make one response available while another remains blocked. This is an interpretation of the event, not a change to the event's measured description.

The same pattern becomes visible elsewhere. Mechanical design, power electronics, fluid transport, computing, physiology, and ecological flows all confront losses of directed availability. The comparison does not claim that the systems share an identical mechanism. It asks whether difference, constraint, pathway, and consequence provide a useful family resemblance across fields that often speak in separate dialects.

For teaching and early research, this distinction can be practical. Ask what remained conserved, what became less recoverable, and which design change could reduce the irreversible route. A learner can state the observation first, the accepted scientific relation second, and the proposed interpretation third. That order reduces the risk that a helpful picture will be remembered as though it were the definition itself.

There is also a limit that should remain audible. Dissipation is not energy destruction. Nor is every distributed motion useless in every context. A conceptual account becomes scientifically stronger only when its terms can be constrained by measurement and when it accepts the possibility of failure. Until then, clarity is valuable, but clarity is not confirmation.

The section therefore returns to the chapter question. The ledger can remain balanced while the architecture becomes less capable of producing the desired directed consequence. In the

Chavanian view, a change is not understood only by naming what changed. We also ask which difference mattered, which route was open, and what new condition the response created.

Pause once more over the ordinary details. Bearings heat, wires resist, fluids mix, shocks form, and turbulence spreads organized motion across many smaller scales. The energy balance can still close while useful performance declines. Now imagine changing only one condition at a time: the material, the size, the surrounding medium, the duration of observation, or the character of the boundary. The visible event might remain recognizable, yet its pace and outcome could alter greatly. This is why a thermodynamic story becomes more accurate when it includes the setting instead of presenting the system as though it acted alone.

Language deserves its own small audit here. The ledger can remain balanced while the architecture becomes less capable of producing the desired directed consequence. In everyday speech we often turn a process into a thing. We say that heat enters, pressure pushes, or equilibrium arrives. Such expressions are efficient, and there is no need to ban them. The useful habit is to ask what the noun stands for in the actual account: a measured relation, an interaction, a redistribution, a constraint, or a shorthand for several of these together.

Cycles and Recreated Delta

Consider what an observer would actually notice. A working machine returns important components to a recurring condition while fuel, electricity, reservoirs, or the environment continually recreate the differences that keep the cycle operating. None of this observation tells us, by itself, which metaphor must be used underneath it. The event gives us readings, timings, boundaries, and consequences. The deeper story is something we add in order to make those relations intelligible.

Thermodynamics and the relevant specialist science already organize the event. A cycle returns the working substance to its initial thermodynamic state, but exchanges with reservoirs and surroundings remain part of the complete process. These established descriptions are the common ground. They allow a teacher in one country, an engineer in another, and a researcher in a third field to speak about the same physical situation without depending on personal imagery.

Now introduce the Chavanian lens carefully. The Chavanian view says the cycle organizes repeated resolution while external architecture restores the Delta required for the next passage. Instead of asking where an invisible content went, the lens asks how the system's condition, boundary, and available degrees of freedom shaped the response. The change of language is useful only if it produces greater discipline rather than a new ambiguity.

This way of speaking travels across disciplines with caution. Industrial cycles, charging systems, seasonal storage, metabolic pathways, and atmospheric circulation all involve recurrence with external support. The recurring architecture may be similar even when the mechanisms, scales, and measurable variables are very different. A membrane is not a piston, a cell is not an engine, and an atmosphere is not a sealed laboratory vessel. Analogy helps only when it does not erase those distinctions.

A teacher can turn this into a conversation. Ask what truly returns, what is consumed or changed outside the cycle, and which conditions must be recreated. Ask which statement came from observation, which came from a tested model, and which came from interpretation. Students often discover that they understood the formal relation but had mixed several physical pictures together. Naming the layers gives them permission to separate those pictures without discarding the science.

The research boundary is equally important. Recurrence is not perpetual motion. Every real cycle includes losses, maintenance, and wider consequences. U E D does not gain experimental status merely because it offers a coherent vocabulary. Structured Vacuum Energy and Alignment Threshold Time require independent definitions, connections to existing measurements, and possible tests. The present book keeps that obligation visible.

What, then, should remain after this section? A repeated path is sustained by a larger system that continually re-establishes the conditions for repetition. The Chavanian view invites another question after the standard explanation is complete: what architecture made this particular thermodynamic response possible?

Let us return to the event without technical haste. A working machine returns important components to a recurring condition while fuel, electricity, reservoirs, or the environment continually recreate the differences that keep the cycle operating. An observer could record what happens, but a fuller explanation would also ask what had to be present before it could happen. There had to be a particular arrangement, a particular contact with the surroundings, and a particular range of possible responses. The event is therefore not only a change in a quantity. It is a change made possible by a physical situation.

This is where ordinary grammar can quietly mislead us. A repeated path is sustained by a larger system that continually re-establishes the conditions for repetition. A convenient sentence may give an abstract quantity the role of an actor, as though the quantity itself travelled, commanded, or chose. Scientists usually understand the shorthand, but new learners may build their intuition around the grammar. Rephrasing the sentence in terms of conditions, interactions, and measurable change often reveals what the original expression had compressed.

Design as Thermodynamic Conversation

A useful way to enter the topic is to slow the event down in imagination. An engineer changes insulation, surface area, flow rate, pressure ratio, material, geometry, timing, or control logic and the same broad system responds differently. At each moment, the system has a condition and a set of possible next steps. Some steps are easy, some require a threshold, and some are excluded by the way the system is built or connected.

The established account tells us how to describe that condition responsibly. Engineering design applies thermodynamics, transport, mechanics, materials, economics, safety, and control to shape measurable performance. It protects us from replacing measurement with intuition. Even in an equation-free conversation, the formal background matters because the words inherit their meaning from that disciplined practice.

From the Chavanian perspective, the event can be narrated as structured resolution. The Chavanian view describes design as deliberate reorganization of resolution architecture within the limits set by physical law. Delta names the relevant difference. The architecture names the permitted and blocked routes. Resolution names the actual process. Consequence names the changed condition that remains after the process has occurred.

That sequence can illuminate another field. The perspective can support design conversations in buildings, batteries, medical devices, agriculture, manufacturing, and climate adaptation. Yet the illumination is conceptual rather than a claim that one universal mechanism has already been demonstrated. The value lies in asking comparable questions: What persists? What differs? What pathway is available? What must the surroundings supply or receive?

In a classroom or research group, the vocabulary can be used as a checking device. Ask students to name which route each design change opens, closes, slows, accelerates, or redirects. If the conversation begins to treat Delta as a force, architecture as a hidden machine, or resolution as a new measurable substance, the terms have drifted. Their purpose is to organize relations, not to populate nature with additional invisible objects.

The strongest caution is this. Conceptual language cannot replace detailed design calculation, standards, testing, or safety analysis. Established thermodynamics can stand without the Chavanian interpretation. The interpretation must therefore earn its place through explanatory usefulness now and through mathematical and experimental work later. It cannot demand immunity from the evidence that shaped the field.

The chapter's central movement appears again. Engineering succeeds by working with the architecture of change, never by negotiating away the laws that constrain it. Change becomes easier to discuss when we keep both the initiating difference and the available pathway in view. The answer is not hidden in either one alone.

Imagine watching the scene with the ability to stop it at several moments. An engineer changes insulation, surface area, flow rate, pressure ratio, material, geometry, timing, or control logic and the same broad system responds differently. At every pause, the system presents both a current condition and a limited range of next conditions. The range is shaped by material structure, contact, geometry, and surrounding constraints. Change is therefore neither a detached command nor a mysterious jump. It occurs through what the present arrangement can physically support.

A small change in wording can expose this structure. Engineering succeeds by working with the architecture of change, never by negotiating away the laws that constrain it. Instead of asking what substance moved, ask which relation changed. Instead of asking what quantity wanted to increase or decrease, ask which processes were permitted under the stated conditions. The older wording may still be retained for convenience, but the second wording makes its hidden assumptions easier to inspect. That is especially helpful when familiar metaphors begin to conflict with one another.

Closing Reflection

Machines make the Chavanian question concrete. They show that the same thermal difference can be allowed to fade, channelled through a cycle, maintained locally, or distributed through losses. The outcomes differ because the architecture differs. The Second Law remains the boundary within which every design must operate.

Machines do not evade thermodynamic direction. They organize the pathways through which Deltas are created, constrained, resolved, and recreated.

Chapter 11

One Architecture, Many Fields

What If Different Fields Are Watching the Same Architecture?

Thermodynamics Arrives Wearing Different Clothes

Thermodynamics rarely arrives alone. In chemistry it comes with reactions and composition. In materials science it comes with phases, defects, and processing. In biology it comes with membranes, metabolism, and regulation. In climate science it comes with radiation, fluids, moisture, and planetary scale. The vocabulary changes, but certain questions return.

The question in this chapter is not whether established thermodynamics works. It is what we notice when we place the idea of architecture beside chemistry, materials, biology, atmosphere, and other disciplinary encounters. The calculation and the measurement remain where science has placed them. Our task is to build a physical narrative that is clear enough for teaching, useful enough for interdisciplinary conversation, and modest enough to preserve the difference between interpretation and evidence.

We will move through familiar scenes, the established account, the Chavanian view, and the limits of that view. By the end, the phrase shared questions can connect fields without erasing their distinct mechanisms should feel less like a slogan and more like a disciplined question that can accompany technical study without competing with it.

Chemistry and Favoured Change

A useful way to enter the topic is to slow the event down in imagination. A reaction mixture contains structures capable of rearrangement. Some changes are thermodynamically favoured under specified conditions, yet the observed result depends on kinetics, pathways, catalysts, and transport. At each moment, the system has a condition and a set of possible next steps. Some steps are easy, some require a threshold, and some are excluded by the way the system is built or connected.

The established account tells us how to describe that condition responsibly. Chemical thermodynamics uses established state functions, chemical potentials, activities, equilibrium criteria, and energy and entropy balances. It protects us from replacing measurement with intuition. Even in an equation-free conversation, the formal background matters because the words inherit their meaning from that disciplined practice.

From the Chavanian perspective, the event can be narrated as structured resolution. The Chavanian view asks how molecular persistence, chemical Delta, accessible mechanism, and changed composition fit into one resolution story. Delta names the relevant difference. The architecture names the permitted and blocked routes. Resolution names the actual process. Consequence names the changed condition that remains after the process has occurred.

That sequence can illuminate another field. The question supports conversations in catalysis, electrochemistry, biochemistry, combustion, corrosion, and atmospheric chemistry. Yet the illumination is conceptual rather than a claim that one universal mechanism has already been demonstrated. The value lies in asking comparable questions: What persists? What differs? What pathway is available? What must the surroundings supply or receive?

In a classroom or research group, the vocabulary can be used as a checking device. Keep thermodynamic favourability, reaction mechanism, and observed rate in separate sentences before connecting them. If the conversation begins to treat Delta as a force, architecture as a hidden machine, or resolution as a new measurable substance, the terms have drifted. Their purpose is to organize relations, not to populate nature with additional invisible objects.

The strongest caution is this. Chemical Delta is not a new measurable replacement for chemical potential or free energy. Established thermodynamics can stand without the Chavanian interpretation. The interpretation must therefore earn its place through explanatory usefulness now and through mathematical and experimental work later. It cannot demand immunity from the evidence that shaped the field.

The chapter's central movement appears again. Chemistry shows that what may happen, how it can happen, and how quickly it happens are different questions. Change becomes easier to discuss when we keep both the initiating difference and the available pathway in view. The answer is not hidden in either one alone.

Imagine watching the scene with the ability to stop it at several moments. A reaction mixture contains structures capable of rearrangement. Some changes are thermodynamically favoured under specified conditions, yet the observed result depends on kinetics, pathways, catalysts, and transport. At every pause, the system presents both a current condition and a limited range of next conditions. The range is shaped by material structure, contact, geometry, and surrounding constraints. Change is therefore neither a detached command nor a mysterious jump. It occurs through what the present arrangement can physically support.

A small change in wording can expose this structure. Chemistry shows that what may happen, how it can happen, and how quickly it happens are different questions. Instead of asking what substance moved, ask which relation changed. Instead of asking what quantity wanted to increase or decrease, ask which processes were permitted under the stated conditions. The older wording may still be retained for convenience, but the second wording makes its hidden assumptions easier to inspect. That is especially helpful when familiar metaphors begin to conflict with one another.

Materials and Layered Response

Begin with the scene itself. A material can conduct, expand, soften, transform, creep, crack, oxidize, or retain memory depending on structure, processing, environment, and scale. It is easy to hurry past such an event because it is familiar. Yet familiarity often hides the number of conditions that must cooperate. The materials have properties. The surroundings have a history. A boundary permits some exchanges and suppresses others. The event looks simple only after all of those details have already been arranged.

Established science gives us the disciplined account. Materials science connects thermodynamics with kinetics, defects, microstructure, mechanics, surfaces, phase diagrams, and processing history. This account does not need a new vocabulary in order to remain valid. It tells us what to measure, which distinctions matter, and how the event can be compared with another event under specified conditions. Any deeper view must begin by respecting that achievement.

The Chavanian question begins one layer lower. The Chavanian view calls this layered accommodation. Different structural levels become available as conditions cross relevant thresholds. The word architecture directs attention away from an imagined travelling substance and toward the relations that make one response available while another remains blocked. This is an interpretation of the event, not a change to the event's measured description.

The same pattern becomes visible elsewhere. Alloys, ceramics, polymers, composites, semiconductors, batteries, and geological materials each display distinct layered architectures. The comparison does not claim that the systems share an identical mechanism. It asks whether difference, constraint, pathway, and consequence provide a useful family resemblance across fields that often speak in separate dialects.

For teaching and early research, this distinction can be practical. Ask which level changed first: electronic, atomic, molecular, microstructural, phase, or macroscopic geometry. A learner can state the observation first, the accepted scientific relation second, and the proposed interpretation third. That order reduces the risk that a helpful picture will be remembered as though it were the definition itself.

There is also a limit that should remain audible. The sequence of layers is material-specific and must be established experimentally. A general hierarchy cannot be imposed in advance. A conceptual account becomes scientifically stronger only when its terms can be constrained by measurement and when it accepts the possibility of failure. Until then, clarity is valuable, but clarity is not confirmation.

The section therefore returns to the chapter question. The same thermal intervention can produce different consequences because materials possess different internal architectures and histories. In the Chavanian view, a change is not understood only by naming what changed. We also ask which difference mattered, which route was open, and what new condition the response created.

Pause once more over the ordinary details. A material can conduct, expand, soften, transform, creep, crack, oxidize, or retain memory depending on structure, processing, environment, and scale. Now imagine changing only one condition at a time: the material, the size, the surrounding medium, the duration of observation, or the character of the boundary. The visible event might remain recognizable, yet its pace and outcome could alter greatly. This is why a thermodynamic story becomes more accurate when it includes the setting instead of presenting the system as though it acted alone.

Language deserves its own small audit here. The same thermal intervention can produce different consequences because materials possess different internal architectures and histories. In everyday speech we often turn a process into a thing. We say that heat enters, pressure pushes, or equilibrium arrives. Such expressions are efficient, and there is no need to ban them. The useful habit is to ask what the noun stands for in the actual account: a measured relation, an interaction, a redistribution, a constraint, or a shorthand for several of these together.

Life as Managed Non-Equilibrium

Consider what an observer would actually notice. A living cell maintains composition, structure, electrical differences, and chemical cycles while exchanging matter and consequence with its surroundings. None of this observation tells us, by itself, which metaphor must be used underneath it. The event gives us readings, timings, boundaries, and consequences. The deeper story is something we add in order to make those relations intelligible.

Thermodynamics and the relevant specialist science already organize the event. Biochemistry, molecular biology, physiology, and non-equilibrium thermodynamics describe metabolism, transport, regulation, replication, and dissipation. These established descriptions are the common ground. They allow a teacher in one country, an engineer in another, and a researcher in a third field to speak about the same physical situation without depending on personal imagery.

Now introduce the Chavanian lens carefully. The Chavanian view sees life as a highly organized architecture for maintaining and resolving many coupled Deltas without claiming that thermodynamics alone explains life. Instead of asking where an invisible content went, the lens asks how the system's condition, boundary, and available degrees of freedom shaped the response. The change of language is useful only if it produces greater discipline rather than a new ambiguity.

This way of speaking travels across disciplines with caution. The perspective may help connect metabolism, homeostasis, ecology, development, and biomedical engineering at the level of constraints and pathways. The recurring architecture may be similar even when the mechanisms, scales, and measurable variables are very different. A membrane is not a piston, a cell is not an engine, and an atmosphere is not a sealed laboratory vessel. Analogy helps only when it does not erase those distinctions.

A teacher can turn this into a conversation. Begin by widening the system boundary and asking what flows make local persistence possible. Ask which statement came from observation, which came from a tested model, and which came from interpretation. Students often discover that they understood the formal relation but had mixed several physical pictures together. Naming the layers gives them permission to separate those pictures without discarding the science.

The research boundary is equally important. Thermodynamic organization is necessary for life but not sufficient to explain biological function, information, evolution, cognition, or agency. U E D does not gain experimental status merely because it offers a coherent vocabulary. Structured Vacuum Energy and Alignment Threshold Time require independent definitions, connections to existing measurements, and possible tests. The present book keeps that obligation visible.

What, then, should remain after this section? Living order persists within thermodynamic law by continuously managing exchanges and producing wider consequence. The Chavanian view invites another question after the standard explanation is complete: what architecture made this particular thermodynamic response possible?

Let us return to the event without technical haste. A living cell maintains composition, structure, electrical differences, and chemical cycles while exchanging matter and consequence with its surroundings. An observer could record what happens, but a fuller explanation would also ask what had to be present before it could happen. There had to be a particular arrangement, a particular contact with the surroundings, and a particular range of possible responses. The event is therefore not only a change in a quantity. It is a change made possible by a physical situation.

This is where ordinary grammar can quietly mislead us. Living order persists within thermodynamic law by continuously managing exchanges and producing wider consequence. A convenient sentence may give an abstract quantity the role of an actor, as though the quantity itself travelled, commanded, or chose. Scientists usually understand the shorthand, but new learners may build their intuition around the grammar. Rephrasing the sentence in terms of conditions, interactions, and measurable change often reveals what the original expression had compressed.

Atmosphere, Ocean, and Planet

A useful way to enter the topic is to slow the event down in imagination. A planet receives radiation, emits radiation, rotates, circulates fluids, moves moisture, changes phase, stores thermal consequence, and presents uneven surfaces and compositions. At each moment, the system has a condition and a set of possible next steps. Some steps are easy, some require a threshold, and some are excluded by the way the system is built or connected.

The established account tells us how to describe that condition responsibly. Climate and weather emerge from radiative transfer, fluid dynamics, thermodynamics, chemistry, geography, biology, and feedback across many scales. It protects us from replacing measurement with intuition. Even in an equation-free conversation, the formal background matters because the words inherit their meaning from that disciplined practice.

From the Chavanian perspective, the event can be narrated as structured resolution. The Chavanian view describes a planetary architecture of persistent and recreated Deltas whose resolutions become winds, currents, clouds, storms, and long-term change. Delta names the relevant difference. The architecture names the permitted and blocked routes. Resolution names the actual process. Consequence names the changed condition that remains after the process has occurred.

That sequence can illuminate another field. The same questions assist communication among atmospheric scientists, oceanographers, geologists, ecologists, and engineers without replacing their models. Yet the illumination is conceptual rather than a claim that one universal mechanism has already been demonstrated. The value lies in asking comparable questions: What persists? What differs? What pathway is available? What must the surroundings supply or receive?

In a classroom or research group, the vocabulary can be used as a checking device. Choose one event, such as cloud formation, and trace the boundaries, phase changes, pressure relations, and wider consequences involved. If the conversation begins to treat Delta as a force, architecture as a hidden machine, or resolution as a new measurable substance, the terms have drifted. Their purpose is to organize relations, not to populate nature with additional invisible objects.

The strongest caution is this. The Chavanian lens is not a climate model and must never be used to dismiss quantitative evidence or uncertainty analysis. Established thermodynamics can stand without the Chavanian interpretation. The interpretation must therefore earn its place through explanatory usefulness now and through mathematical and experimental work later. It cannot demand immunity from the evidence that shaped the field.

The chapter's central movement appears again. Planetary change is thermodynamic architecture operating through coupled fields, materials, and timescales. Change becomes easier to discuss when we keep both the initiating difference and the available pathway in view. The answer is not hidden in either one alone.

Imagine watching the scene with the ability to stop it at several moments. A planet receives radiation, emits radiation, rotates, circulates fluids, moves moisture, changes phase, stores thermal consequence, and presents uneven surfaces and compositions. At every pause, the system presents both a current condition and a limited range of next conditions. The range is shaped by material structure, contact, geometry, and surrounding constraints. Change is therefore neither a detached command nor a mysterious jump. It occurs through what the present arrangement can physically support.

A small change in wording can expose this structure. Planetary change is thermodynamic architecture operating through coupled fields, materials, and timescales. Instead of asking what substance moved, ask which relation changed. Instead of asking what quantity wanted to increase or decrease, ask which processes were permitted under the stated conditions. The older wording may still be retained for convenience, but the second wording makes its hidden assumptions easier to inspect. That is especially helpful when familiar metaphors begin to conflict with one another.

Stars, Gravity, and the Edge of the View

Begin with the scene itself. Stars, accretion flows, expanding gases, radiation fields, compact objects, and the universe as a whole invite thermodynamic language at scales far beyond the laboratory. It is easy to hurry past such an event because it is familiar. Yet familiarity often hides the number of conditions that must cooperate. The materials have properties. The surroundings have a history. A boundary permits some exchanges and suppresses others. The event looks simple only after all of those details have already been arranged.

Established science gives us the disciplined account. Astrophysical and cosmological thermodynamics must be joined with gravitation, relativity, radiation, quantum physics, plasma physics, and observational evidence. This account does not need a new vocabulary in order to remain valid. It tells us what to measure, which distinctions matter, and how the event can be compared with another event under specified conditions. Any deeper view must begin by respecting that achievement.

The Chavanian question begins one layer lower. The Chavanian framework asks whether persistence and gradient within a continuous foundation might eventually connect thermodynamics with wider physical architecture. The word architecture directs attention away from an imagined travelling substance and toward the relations that make one response available while another remains blocked. This is an interpretation of the event, not a change to the event's measured description.

The same pattern becomes visible elsewhere. The question relates to other Chavanian work on gravity and continuity, but it remains a frontier rather than a conclusion of this companion. The comparison does not claim that the systems share an identical mechanism. It asks whether difference, constraint, pathway, and consequence provide a useful family resemblance across fields that often speak in separate dialects.

For teaching and early research, this distinction can be practical. Use cosmic examples to demonstrate how quickly an intuitive thermodynamic picture reaches the limits of its domain. A learner can state the observation first, the accepted scientific relation second, and the proposed interpretation third. That order reduces the risk that a helpful picture will be remembered as though it were the definition itself.

There is also a limit that should remain audible. Cosmological extension carries the highest evidential burden. U E D must not borrow authority from established cosmology without reproducing its mathematical and observational successes. A conceptual account becomes scientifically stronger only when its terms can be constrained by measurement and when it accepts the possibility of failure. Until then, clarity is valuable, but clarity is not confirmation.

The section therefore returns to the chapter question. A view may suggest new connections, but the farther it travels from tested domains, the more carefully it must distinguish invitation from evidence. In the Chavanian view, a change is not understood only by naming what changed. We also ask which difference mattered, which route was open, and what new condition the response created.

Pause once more over the ordinary details. Stars, accretion flows, expanding gases, radiation fields, compact objects, and the universe as a whole invite thermodynamic language at scales far beyond the laboratory. Now imagine changing only one condition at a time: the material, the size, the surrounding medium, the duration of observation, or the character of the boundary. The visible event might remain recognizable, yet its pace and outcome could alter greatly. This is why a thermodynamic story becomes more accurate when it includes the setting instead of presenting the system as though it acted alone.

Language deserves its own small audit here. A view may suggest new connections, but the farther it travels from tested domains, the more carefully it must distinguish invitation from evidence. In everyday speech we often turn a process into a thing. We say that heat enters, pressure pushes, or equilibrium arrives. Such expressions are efficient, and there is no need to ban them. The useful habit is to ask what the noun stands for in the actual account: a measured relation, an interaction, a redistribution, a constraint, or a shorthand for several of these together.

Closing Reflection

Thermodynamic architecture can serve as a conversational bridge across fields because difference, boundary, pathway, persistence, and consequence recur widely. The bridge is useful only when it carries the specific science of each field with it. A membrane is not a turbine, a cell is not a planet, and a common vocabulary is not a common mechanism.

Different fields may encounter comparable architectures of change while retaining distinct variables, mechanisms, scales, and standards of evidence.

Chapter 12

From View to Science

What If a View Wants to Become Science?

An Elegant Story Is Only the Beginning

A conceptual view can be useful before it is a theory. It can expose hidden assumptions, connect questions, and improve teaching. But science asks for more. Terms must become operational. Relations must become quantitative. Known results must be recovered. New claims must risk being wrong.

The question in this chapter is not whether established thermodynamics works. It is what we notice when we place the idea of architecture beside definition, formalization, recovery, prediction, evidence, and falsifiability. The calculation and the measurement remain where science has placed them. Our task is to build a physical narrative that is clear enough for teaching, useful enough for interdisciplinary conversation, and modest enough to preserve the difference between interpretation and evidence.

We will move through familiar scenes, the established account, the Chavanian view, and the limits of that view. By the end, the phrase the Chavanian view must become constrained and vulnerable if it is to move beyond interpretation should feel less like a slogan and more like a disciplined question that can accompany technical study without competing with it.

From Vocabulary to Definition

Begin with the scene itself. Words such as Delta, resolution, architecture, persistence, and consequence can feel clear in examples while remaining too broad for calculation. It is easy to hurry past such an event because it is familiar. Yet familiarity often hides the number of conditions that must cooperate. The materials have properties. The surroundings have a history. A boundary permits some exchanges and suppresses others. The event looks simple only after all of those details have already been arranged.

Established science gives us the disciplined account. Scientific concepts gain strength through operational definitions, measurable variables, reproducible procedures, dimensional consistency, and constrained domains. This account does not need a new vocabulary in order to remain valid. It tells us what to measure, which distinctions matter, and how the event can be compared with another event under specified conditions. Any deeper view must begin by respecting that achievement.

The Chavanian question begins one layer lower. The Chavanian programme must specify which Deltas are physical variables, how resolution geometry is represented, and what Alignment Threshold Time adds beyond established kinetics. The word architecture directs attention away from an imagined travelling substance and toward the relations that make one response available while another remains blocked. This is an interpretation of the event, not a change to the event's measured description.

The same pattern becomes visible elsewhere. The same transition from qualitative concept to quantitative discipline appears throughout the history of temperature, energy, fields, genes, and information. The comparison does not claim that the systems share an identical mechanism. It asks whether difference, constraint, pathway, and consequence provide a useful family resemblance across fields that often speak in separate dialects.

For teaching and early research, this distinction can be practical. Ask students to turn one appealing sentence into a measurement proposal and notice where ambiguity appears. A learner can state the observation first, the accepted scientific relation second, and the proposed interpretation third. That order reduces the risk that a helpful picture will be remembered as though it were the definition itself.

There is also a limit that should remain audible. A term does not become scientific because it is capitalized or repeated consistently. A conceptual account becomes scientifically stronger only when its terms can be constrained by measurement and when it accepts the possibility of failure. Until then, clarity is valuable, but clarity is not confirmation.

The section therefore returns to the chapter question. The first research obligation is to make the vocabulary precise enough that two independent investigators would use it in the same way. In the Chavanian view, a change is not understood only by naming what changed. We also ask which difference mattered, which route was open, and what new condition the response created.

Pause once more over the ordinary details. Words such as Delta, resolution, architecture, persistence, and consequence can feel clear in examples while remaining too broad for calculation. Now imagine changing only one condition at a time: the material, the size, the surrounding medium, the duration of observation, or the character of the boundary. The visible event might remain recognizable, yet its pace and outcome could alter greatly. This is why a thermodynamic story becomes more accurate when it includes the setting instead of presenting the system as though it acted alone.

Language deserves its own small audit here. The first research obligation is to make the vocabulary precise enough that two independent investigators would use it in the same way. In everyday speech we often turn a process into a thing. We say that heat enters, pressure pushes, or equilibrium arrives. Such expressions are efficient, and there is no need to ban them. The useful habit is to ask what the noun stands for in the actual account: a measured relation, an interaction, a redistribution, a constraint, or a shorthand for several of these together.

Recovery Before Replacement

Consider what an observer would actually notice. Thermodynamics, statistical mechanics, transport theory, and quantum physics already explain and predict a vast range of observations. None of this observation tells us, by itself, which metaphor must be used underneath it. The event gives us readings, timings, boundaries, and consequences. The deeper story is something we add in order to make those relations intelligible.

Thermodynamics and the relevant specialist science already organize the event. Any deeper framework must reproduce established results in the domains where those results have been repeatedly confirmed. These established descriptions are the common ground. They allow a teacher in one country, an engineer in another, and a researcher in a third field to speak about the same physical situation without depending on personal imagery.

Now introduce the Chavanian lens carefully. U E D must show how familiar thermodynamic quantities and laws emerge from its proposed continuity, not simply place new words beneath old equations. Instead of asking where an invisible content went, the lens asks how the system's condition, boundary, and available degrees of freedom shaped the response. The change of language is useful only if it produces greater discipline rather than a new ambiguity.

This way of speaking travels across disciplines with caution. Successful foundational theories are judged partly by whether they recover the practical structures that preceded them. The recurring architecture may be similar even when the mechanisms, scales, and measurable variables are very different. A membrane is not a piston, a cell is not an engine, and an atmosphere is not a sealed laboratory vessel. Analogy helps only when it does not erase those distinctions.

A teacher can turn this into a conversation. Make a list of results the proposed framework is not allowed to change merely for convenience. Ask which statement came from observation, which came from a tested model, and which came from interpretation. Students often discover that they understood the formal relation but had mixed several physical pictures together. Naming the layers gives them permission to separate those pictures without discarding the science.

The research boundary is equally important. Reinterpretation without recovery remains philosophy. Claiming replacement without recovery is scientifically premature. U E D does not gain experimental status merely because it offers a coherent vocabulary. Structured Vacuum Energy and Alignment Threshold Time require independent definitions, connections to existing measurements, and possible tests. The present book keeps that obligation visible.

What, then, should remain after this section? The equations do not owe the new view an explanation. The new view owes the equations a derivation or a demonstrably compatible foundation. The Chavanian view invites another question after the standard explanation is complete: what architecture made this particular thermodynamic response possible?

Let us return to the event without technical haste. Thermodynamics, statistical mechanics, transport theory, and quantum physics already explain and predict a vast range of observations. An observer could record what happens, but a fuller explanation would also ask what had to be present before it could happen. There had to be a particular arrangement, a particular contact with the surroundings, and a particular range of possible responses. The event is therefore not only a change in a quantity. It is a change made possible by a physical situation.

This is where ordinary grammar can quietly mislead us. The equations do not owe the new view an explanation. The new view owes the equations a derivation or a demonstrably compatible foundation. A convenient sentence may give an abstract quantity the role of an actor, as though the quantity itself travelled, commanded, or chose. Scientists usually understand the shorthand, but new learners may build their intuition around the grammar. Rephrasing the sentence in terms of conditions, interactions, and measurable change often reveals what the original expression had compressed.

What Would Count as a Prediction?

A useful way to enter the topic is to slow the event down in imagination. A framework may retell every known event in its own vocabulary and still make no distinguishable claim about what will be observed next. At each moment, the system has a condition and a set of possible next steps. Some steps are easy, some require a threshold, and some are excluded by the way the system is built or connected.

The established account tells us how to describe that condition responsibly. A scientific prediction specifies conditions, expected outcomes, uncertainty, and a contrast with alternatives that can be tested. It protects us from replacing measurement with intuition. Even in an equation-free conversation, the formal background matters because the words inherit their meaning from that disciplined practice.

From the Chavanian perspective, the event can be narrated as structured resolution. The Chavanian programme must identify a domain in which U E D, resolution geometry, or Alignment Threshold Time predicts a measurable difference not inserted after the result is known. Delta names the relevant difference. The architecture names the permitted and blocked routes. Resolution names the actual process. Consequence names the changed condition that remains after the process has occurred.

That sequence can illuminate another field. Non-equilibrium processes, interfacial phenomena, threshold behaviour, and cross-scale dynamics may offer research domains, but only careful formal work can decide. Yet the illumination is conceptual rather than a claim that one universal mechanism has already been demonstrated. The value lies in asking comparable questions: What persists? What differs? What pathway is available? What must the surroundings supply or receive?

In a classroom or research group, the vocabulary can be used as a checking device. Ask what observation would be different if the framework were false and whether the proposed test can distinguish the interpretations. If the conversation begins to treat Delta as a force, architecture as a hidden machine, or resolution as a new measurable substance, the terms have drifted. Their purpose is to organize relations, not to populate nature with additional invisible objects.

The strongest caution is this. A new description of an old result is not a new prediction. Established thermodynamics can stand without the Chavanian interpretation. The interpretation must therefore earn its place through explanatory usefulness now and through mathematical and experimental work later. It cannot demand immunity from the evidence that shaped the field.

The chapter's central movement appears again. Prediction begins when the framework accepts a measurable risk that established alternatives do not share. Change becomes easier to discuss when we keep both the initiating difference and the available pathway in view. The answer is not hidden in either one alone.

Imagine watching the scene with the ability to stop it at several moments. A framework may retell every known event in its own vocabulary and still make no distinguishable claim about what will be observed next. At every pause, the system presents both a current condition and a limited range of next conditions. The range is shaped by material structure, contact, geometry, and surrounding constraints. Change is therefore neither a detached command nor a mysterious jump. It occurs through what the present arrangement can physically support.

A small change in wording can expose this structure. Prediction begins when the framework accepts a measurable risk that established alternatives do not share. Instead of asking what substance moved, ask which relation changed. Instead of asking what quantity wanted to increase or decrease, ask which processes were permitted under the stated conditions. The older wording may still be retained for convenience, but the second wording makes its hidden assumptions easier to inspect. That is especially helpful when familiar metaphors begin to conflict with one another.

Falsifiability and Revision

Begin with the scene itself. Researchers naturally become attached to structures they have developed. A vocabulary can become an identity, making contrary evidence feel like a personal threat. It is easy to hurry past such an event because it is familiar. Yet familiarity often hides the number of conditions that must cooperate. The materials have properties. The surroundings have a history. A boundary permits some exchanges and suppresses others. The event looks simple only after all of those details have already been arranged.

Established science gives us the disciplined account. Scientific practice relies on reproducibility, criticism, comparison, uncertainty, peer engagement, and revision in response to evidence. This account does not need a new vocabulary in order to remain valid. It tells us what to measure, which distinctions matter, and how the event can be compared with another event under specified conditions. Any deeper view must begin by respecting that achievement.

The Chavanian question begins one layer lower. The Chavanian view includes a Shared-Reality Principle: established observations cannot be altered or dismissed merely to protect the interpretation. The word architecture directs attention away from an imagined travelling substance and toward the relations that make one response available while another remains blocked. This is an interpretation of the event, not a change to the event's measured description.

The same pattern becomes visible elsewhere. The principle applies to every speculative programme, independent research project, and interdisciplinary synthesis. The comparison does not claim that the systems share an identical mechanism. It asks whether difference, constraint, pathway, and consequence provide a useful family resemblance across fields that often speak in separate dialects.

For teaching and early research, this distinction can be practical. Invite readers to write down in advance what result would weaken, revise, or falsify a proposed claim. A learner can state the observation first, the accepted scientific relation second, and the proposed interpretation third. That order reduces the risk that a helpful picture will be remembered as though it were the definition itself.

There is also a limit that should remain audible. Falsifiability is not achieved by adding a sentence that the theory could be wrong. The claim must be specific enough to encounter evidence. A conceptual account becomes scientifically stronger only when its terms can be constrained by measurement and when it accepts the possibility of failure. Until then, clarity is valuable, but clarity is not confirmation.

The section therefore returns to the chapter question. A framework becomes scientifically mature when it can learn from failure instead of redefining every failure as hidden success. In the Chavanian view, a change is not understood only by naming what changed. We also ask which difference mattered, which route was open, and what new condition the response created.

Pause once more over the ordinary details. Researchers naturally become attached to structures they have developed. A vocabulary can become an identity, making contrary evidence feel like a personal threat. Now imagine changing only one condition at a time: the material, the size, the surrounding medium, the duration of observation, or the character of the boundary. The visible event might remain recognizable, yet its pace and outcome could alter greatly. This is why a thermodynamic story becomes more accurate when it includes the setting instead of presenting the system as though it acted alone.

Language deserves its own small audit here. A framework becomes scientifically mature when it can learn from failure instead of redefining every failure as hidden success. In everyday speech we often turn a process into a thing. We say that heat enters, pressure pushes, or equilibrium arrives. Such expressions are efficient, and there is no need to ban them. The useful habit is to ask what the noun stands for in the actual account: a measured relation, an interaction, a redistribution, a constraint, or a shorthand for several of these together.

The Value of a View Before Theory

Consider what an observer would actually notice. Not every useful idea begins as a complete quantitative theory. Conceptual maps can improve questions, reveal category errors, and support communication across disciplines. None of this observation tells us, by itself, which metaphor must be used underneath it. The event gives us readings, timings, boundaries, and consequences. The deeper story is something we add in order to make those relations intelligible.

Thermodynamics and the relevant specialist science already organize the event. Science includes exploratory models, interpretations, heuristics, and research programmes alongside mature predictive theories, provided their status is stated honestly. These established descriptions are the common ground. They allow a teacher in one country, an engineer in another, and a researcher in a third field to speak about the same physical situation without depending on personal imagery.

Now introduce the Chavanian lens carefully. The present Chavanian contribution is a disciplined view of difference, architecture, and resolution that may guide later formalization. Instead of asking where an invisible content went, the lens asks how the system's condition, boundary, and available degrees of freedom shaped the response. The change of language is useful only if it produces greater discipline rather than a new ambiguity.

This way of speaking travels across disciplines with caution. Teachers can use the view to organize discussion, and researchers can use it to generate questions, without treating it as established fact. The recurring architecture may be similar even when the mechanisms, scales, and measurable variables are very different. A membrane is not a piston, a cell is not an engine, and an atmosphere is not a sealed laboratory vessel. Analogy helps only when it does not erase those distinctions.

A teacher can turn this into a conversation. End every application by naming what was clarified and what remains unmeasured. Ask which statement came from observation, which came from a tested model, and which came from interpretation. Students often discover that they understood the formal relation but had mixed several physical pictures together. Naming the layers gives them permission to separate those pictures without discarding the science.

The research boundary is equally important. Explanatory satisfaction can be mistaken for truth. A story that feels unified may still be incomplete or wrong. U E D does not gain experimental status merely because it offers a coherent vocabulary. Structured Vacuum Energy and Alignment Threshold Time require independent definitions, connections to existing measurements, and possible tests. The present book keeps that obligation visible.

What, then, should remain after this section? A view earns present value through clarity and future value through the research discipline it is willing to accept. The Chavanian view invites another question after the standard explanation is complete: what architecture made this particular thermodynamic response possible?

Let us return to the event without technical haste. Not every useful idea begins as a complete quantitative theory. Conceptual maps can improve questions, reveal category errors, and support communication across disciplines. An observer could record what happens, but a fuller explanation would also ask what had to be present before it could happen. There had to be a particular arrangement, a particular contact with the surroundings, and a particular range of possible responses. The event is therefore not only a change in a quantity. It is a change made possible by a physical situation.

This is where ordinary grammar can quietly mislead us. A view earns present value through clarity and future value through the research discipline it is willing to accept. A convenient sentence may give an abstract quantity the role of an actor, as though the quantity itself travelled, commanded, or chose. Scientists usually understand the shorthand, but new learners may build their intuition around the grammar. Rephrasing the sentence in terms of conditions, interactions, and measurable change often reveals what the original expression had compressed.

Closing Reflection

The Chavanian view should not be protected from thermodynamics, statistical mechanics, or experiment. It should be built to meet them. Its immediate role is conversational and interpretive. Its future depends on definition, mathematics, recovery, prediction, and test. That path is demanding, but anything less would leave the view permanently beneath the threshold of physical theory.

The Chavanian view begins as an organized interpretation. It can approach theory only by becoming mathematically precise, empirically distinguishable, and open to rejection.

Closing Reflection

The Equations Can Stay Where They Are

We began with a cup cooling on a desk and ended with a research programme asking to become vulnerable to experiment. Between those points, the book passed through heat, temperature, pressure, equilibrium, persistence, boundaries, entropy, machines, living systems, materials, atmosphere, and the proposed foundation of Universal Energy Dynamics.

The journey did not require an equation because its task was not calculation. It was to organize a way of looking. Technical readers will know that the equations remain essential. Non-technical readers should leave with respect for that necessity. A conceptual narrative is successful only when it helps the formal structure become clearer, not when it encourages the reader to ignore it.

The central question was whether change has an architecture. The answer offered by the Chavanian view is yes, provided architecture is understood carefully. Every physical event occurs within a system whose boundaries, material properties, degrees of freedom, couplings, thresholds, history, and surroundings make some responses available and others unavailable.

Difference begins the question. The book used the word Delta for a relational difference relevant to possible change.

Delta was never given agency. It did not push, decide, seek, or command. It altered the condition from which a physically available process could occur.

Boundary shaped the question. A wall, piston, membrane, interface, open window, valve, or analytical system boundary determined what could cross, what could move, what could be counted, and which processes became internal or external to the description.

Pathway shaped the event. A thermodynamically favoured change could wait behind a kinetic barrier. A catalyst could open a faster route. Insulation could delay thermal exchange. A machine could organize a cycle. A living system could spend resources to maintain local difference.

Resolution named the process. It did not mean that every difference vanished. It meant that a particular relevant relation changed through an available physical route. The process could reduce one Delta while producing another, export consequence to the surroundings, or reorganize the system so that different responses became possible afterward.

Consequence named the changed condition. A cup cooled and the room warmed slightly. A gas expanded and the boundary moved. A reaction altered composition. A phase transition changed structure. A refrigerator preserved a cool compartment while warming the wider environment.

Persistence kept the world from becoming a story only about disappearance. Systems can retain identity while participating in ongoing exchange. Equilibrium can remove a particular net tendency while structure remains. Steady states can maintain readings through continuous throughput. Matter can be treated as real and persistent without assuming that its deepest boundary is absolute separation.

Entropy kept the view disciplined. The book did not replace entropy with distributed consequence. It retained entropy as an established thermodynamic state quantity and asked whether distribution provides a useful physical interpretation of irreversibility and reduced accessibility of directed return.

Machines demonstrated the architecture. Engines channel part of a thermal process into work while rejecting consequence elsewhere. Refrigerators spend work to maintain local thermal difference. Dissipation preserves the energy ledger while reducing useful direction. Design succeeds by reorganizing pathways within law.

Different fields showed both the reach and the danger of the view. Chemistry, materials, biology, atmosphere, and astrophysics all encounter differences, constraints, pathways, and persistence. Yet each field has its own variables, mechanisms, models, scales, and evidence. Shared questions do not prove a shared fundamental mechanism.

Universal Energy Dynamics entered as a proposal beneath the interpretive layer. It suggested a continuous foundational reality called Structured Vacuum Energy and treated matter as persistent structured manifestation within that continuity. Alignment Threshold Time appeared as a proposed account of accessibility. None of those proposals was presented as established thermodynamics.

This reading boundary is the ethical centre of the book. Observation belongs to nature. Established relations belong to the scientific tradition that tested them. The Chavanian view belongs to interpretation. U E D belongs to a developing foundational proposal. Future prediction, if it becomes possible, must belong to experiment.

A teacher may use the view to ask better questions. A student may use it to separate a definition from a metaphor. An early researcher may use it to notice an assumption that deserves operational clarification. A scientist from another field may use it to connect thermodynamic themes without abandoning disciplinary precision.

The equations can therefore stay where they are. The laws can stay where they are. The measurements can stay where they are. The Chavanian view asks only whether another physical narrative can sit beneath them and become useful without pretending to possess authority it has not yet earned.

What if change has an architecture? The answer is not a final theory. It is a disciplined invitation: look for what persists, identify what differs, state the boundary, find the available pathway, follow the consequence, and then ask what the changed condition makes possible next.

Core Definitions

A Listening Guide to the Chavanian Vocabulary

The definitions in this section are intentionally verbal. They are meant to help a listener keep the levels of the book distinct. They are not substitutes for technical thermodynamic definitions.

Architecture means the organized pattern of boundaries, material properties, degrees of freedom, couplings, constraints, thresholds, histories, and possible pathways through which a system can respond.

Boundary means both the physical interface through which interactions occur and the analytical choice that determines what is included in the system. A boundary can permit different forms of exchange in different ways.

Changed geometry means the new arrangement of relations after a process. Geometry is used broadly here and can include physical shape, connectivity, phase, composition, constraint, and available pathway.

Chavanian view means the proposed interpretation that organizes events around persistence, Delta, boundary, accessibility, resolution, consequence, and changed geometry while preserving established scientific observations and relations.

Consequence means the changed physical condition produced through a process. It may appear as altered temperature, pressure, volume, composition, phase, motion, distribution, structure, or surroundings.

Delta means a relational difference relevant to what a defined system can do. Delta is not a force, substance, energy, or agent. Technical variables retain their own definitions.

Distributed consequence means the interpretive idea that an initially directed or localized condition can become spread across more degrees of freedom, microscopic configurations, pathways, or surroundings. It is not a replacement definition for entropy.

Equilibrium means the established thermodynamic condition appropriate to the variables and constraints of the system. In Chavanian language, the relevant Deltas no longer produce net change through the available pathways under those conditions.

Established science means observations, definitions, relations, models, and theories supported within their demonstrated domains by measurement, experiment, mathematical coherence, and scientific practice.

Heat remains the established process quantity associated with energy transfer across a boundary because of temperature difference. The Chavanian view interprets the event as thermal resolution rather than the movement of a material called heat.

Interpretation means a physical or conceptual account of what established descriptions may correspond to beneath their successful operational use. An interpretation can be valuable without being experimentally distinct.

Manifestation means an identifiable structured condition that persists within a wider physical reality. The term does not deny objects, matter, molecules, or measurable systems.

Pathway means the actual physical route through which change can occur. A pathway may involve transport, reaction, boundary motion, phase transformation, radiation, diffusion, or another domain-specific process.

Persistence means the continued identity or viability of a structure or condition across a relevant timescale despite internal activity and external interaction.

Pressure remains the established thermodynamic and mechanical quantity. The Chavanian view emphasizes its expression through the constrained relation between a system and the boundary or field in which it is measured.

Resolution means the physical process through which a relevant Delta diminishes, transforms, redistributes, creates consequence, or becomes replaced within a changed condition. Resolution does not imply the disappearance of all difference.

Resolution geometry means the proposed organization of permitted and blocked pathways through which a Delta can become consequence. It is presently a conceptual term requiring formalization.

Steady state means a condition in which selected macroscopic variables remain approximately stable through balanced flows or processes. It should not be confused with thermodynamic equilibrium.

Structured Vacuum Energy means the proposed continuous foundational substrate of Universal Energy Dynamics. It is not an established thermodynamic substance, not a synonym for the accepted quantum vacuum, and not presently a complete predictive theory.

Temperature remains the established thermodynamic coordinate that supports comparison of thermal states and equilibrium. The Chavanian view treats it as a reading of condition rather than a material content.

Universal Energy Dynamics means the wider proposed foundational framework in which continuity, Structured Vacuum Energy, persistence, Delta, and resolution are developed. In this book it is presented as provisional.

Alignment Threshold Time means a proposed Chavanian concept associated with when a pathway becomes physically accessible. It must not be confused with established activation energies, critical points, relaxation times, or kinetic parameters.

The Shared-Reality Principle means that the Chavanian interpretation and conventional science must confront the same physical observations. No empirical result may be altered, ignored, or redefined merely to protect the framework.

A Suggested Reading Path

For Teachers, Students, Researchers, and Scientists Across Fields

Teachers may read the Opening Note, Introduction, and the first six chapters before deciding which later chapters match their course. Each chapter can support one discussion period. Begin with the familiar scene, ask students for ordinary explanations, and then separate observation, established relation, and interpretation.

Students should use the book beside a formal thermodynamics text. When the companion makes a concept feel intuitive, return to the technical definition and test whether the intuition agrees. When they conflict, preserve the technical definition and revise the intuition.

Early researchers may begin with Chapters Eight, Nine, Eleven, and Twelve. Those chapters distinguish thermodynamic tendency from pathway, entropy from metaphor, interdisciplinary analogy from mechanism, and interpretation from prediction.

Chemists may focus on phase change, equilibrium, pathways, entropy, and the chapter that crosses fields. Materials scientists may focus on boundaries, pressure, persistence, layered response, thresholds, history, and dissipation.

Biologists and medical readers may focus on maintained Deltas, selective boundaries, temperature regulation, steady states, entropy, and the discussion of life as managed non-equilibrium.

Engineers may move from the chapters on heat and boundary directly to machines, then return to temperature, pressure, entropy, and research discipline. The Chavanian language is most useful when it clarifies design constraints without replacing calculation.

Atmospheric, ocean, Earth, and planetary scientists may begin with Delta, boundary, local equilibrium, pressure, entropy, and the cross-disciplinary chapter. They should treat the architecture as a communication aid rather than a forecasting model.

Readers interested in Universal Energy Dynamics should not skip the earlier chapters. The U E D layer is safest when entered after the established quantities and system boundaries are clear. Continuity should be understood as a proposal beneath thermodynamics, not as a shortcut around it.

Listeners may pause after each Closing Reflection and restate three sentences aloud: what established science says, what the Chavanian view adds, and what remains provisional. If those sentences cannot be separated, the chapter should be heard again before moving forward.

Selected Further Reading

Established Thermodynamics and Thermal Physics

This companion contains no equations and is not designed to provide technical training. Readers who need formal thermodynamics should study established textbooks, laboratory methods, and the specialist literature of their field.

Herbert Callen's Thermodynamics and an Introduction to Thermostatistics offers a rigorous treatment of thermodynamic structure and the role of state principles.

Enrico Fermi's Thermodynamics remains a concise classical presentation of the subject's foundational relations.

Daniel Schroeder's An Introduction to Thermal Physics provides an accessible route through thermodynamics and statistical mechanics for students with the necessary mathematical background.

Frederick Reif's Fundamentals of Statistical and Thermal Physics develops the statistical foundations in depth.

Charles Kittel and Herbert Kroemer's Thermal Physics offers a compact and influential treatment of thermodynamic and statistical reasoning.

Mark Zemansky and Richard Dittman's Heat and Thermodynamics provides a broad classical account with attention to physical systems and measurement.

Peter Atkins and Julio de Paula's Physical Chemistry connects thermodynamics with chemistry, phase behaviour, equilibrium, and molecular interpretation.

Ilya Prigogine's work on irreversible processes provides historical and conceptual background for non-equilibrium thermodynamics.

Readers working in materials, biology, atmosphere, engineering, information, or cosmology should pair these foundations with current specialist texts and peer-reviewed literature in their own domain.

The purpose of this list is not to suggest that one text settles every interpretive question. It identifies part of the established scientific ground that any Chavanian development must respect and eventually recover.

Acknowledgements

The Scientific Ground Beneath the Conversation

This book depends first on the generations of scientists, engineers, teachers, technicians, and students who built and tested thermodynamics. The subject became durable because measurement, definition, experiment, and calculation repeatedly corrected intuition.

I acknowledge the foundational work associated with Carnot, Clausius, Kelvin, Boltzmann, Gibbs, Planck, Nernst, Einstein, Debye, and many others whose contributions shaped the language and practice of thermal science.

I am grateful to teachers who know that a correct calculation and a clear physical explanation are related but different achievements. Their questions often reveal where familiar language has become more literal than the science requires.

I am equally grateful to students who continue asking where heat goes, what pressure is, why equilibrium permits motion, why entropy increases, and how living systems remain organized. Such questions are not signs of weakness. They are reminders that understanding has layers.

Researchers and scientists across fields also shaped the need for this companion. Thermodynamics appears in chemistry, materials, biology, medicine, atmosphere, Earth science, computing, and engineering, yet each field inherits a different vocabulary and set of intuitions.

The Chavanian framework itself developed through continuing work on energy, gravity, perception, cognition, intelligence, persistence, Delta, and resolution. Bringing those ideas into thermodynamics has forced clearer distinctions between observation, measurement, interpretation, and proposal.

The speculative elements remain unfinished. Structured Vacuum Energy, Alignment Threshold Time, and deeper resolution geometry require mathematical and experimental development. Criticism that clarifies those obligations is part of the work, not an interruption of it.

Finally, I acknowledge the reader who is willing to hold two commitments at once: respect for established science and openness to foundational questions. This book exists for that reader.

About the Author

Sandeep J. Chavan is an independent researcher, educator, and author whose work explores foundational questions across physics, cognition, intelligence, perception, identity, artificial intelligence, and complex human systems.

His research and writing are organized around a growing conceptual body of work sometimes described as the Chavanian ecosystem. Across this work he develops Universal Energy Dynamics, the Universal Theory of Resolution, Alignment Threshold Time, Structured Vacuum Energy, and broader ideas involving Delta, persistence, manifestation, consequence, and resolution geometry.

Chavan's method generally begins without asking successful equations to surrender their demonstrated role. It asks what physical or conceptual architecture might exist beneath descriptions that already work.

A recurring theme is the distinction between description and ontology. Mathematics may describe relations with extraordinary precision without necessarily settling what fundamentally exists beneath those relations. An established model can remain valid within its demonstrated domain while foundational interpretation remains open.

Chavan's background crosses technical, industrial, and educational environments. Work in engineering and industry, followed by sustained engagement with science and mathematics education, shaped a writing style that seeks accessibility without treating conceptual questions as slogans.

His books repeatedly emphasize that coherence is not confirmation. Any proposed deeper physical framework must eventually become mathematical, predictive, and falsifiable if it is to move from interpretation toward physical theory.

Across the wider project, diverse subjects are connected by a common question: how does difference become consequence within a structured physical or cognitive system?

This companion applies that question to thermodynamics while deliberately narrowing the claim. The laws and measurable quantities remain intact. The Chavanian view is offered as another window through which their physical relationships may be discussed.

Further Reading Within the Chavanian Ecosystem

A Path Beyond This Companion

Chavanian Thermodynamics is the detailed companion to the present conversational volume. It develops the thermodynamic argument at much greater length and includes the formal concepts, distinctions, equations, appendices, and research boundaries omitted here.

Rethinking Energy examines the tendency to imagine energy as a substance that is stored, transported, consumed, or converted into different material forms. It provides an earlier route into the conceptual questions that later shaped the thermodynamic work.

Universal Theory of Resolution develops the broad movement from difference through resolution, manifestation, consequence, and changed geometry across domains beyond thermodynamics.

Gravity: A Consequence of Gradient explores the proposed relation between persistence, continuity, gradient, and gravitation within the wider U E D programme.

More General Relativity continues the attempt to connect continuity and relational physical architecture with questions traditionally addressed through gravitation and relativity.

Other works in the Chavanian ecosystem apply related distinctions to perception, cognition, intelligence, identity, operational dynamics, and artificial systems. These books are not prerequisites for thermodynamics and should not be treated as evidence for the physical proposals.

Readers interested mainly in established thermal science should begin with standard thermodynamics and statistical mechanics. Readers interested in the interpretive layer may then approach the Chavanian works in the order that matches their questions.

Disclaimer and Statement of Scope

Where the Book Must Stop

This book is a conceptual and educational companion. It is not a replacement for formal thermodynamics, statistical mechanics, kinetic theory, transport theory, quantum physics, chemistry, materials science, biology, engineering design, climate science, or any other established specialist discipline.

The absence of equations is a design choice for accessibility and narration. It does not imply that thermodynamic questions can be resolved without mathematics. Readers must use appropriate technical sources for calculation, experimental design, safety, engineering, medicine, or policy.

The terms temperature, pressure, heat, work, internal energy, entropy, equilibrium, phase, chemical potential, and related quantities retain their established scientific meanings. Conversational explanations in this book do not redefine those quantities.

The Chavanian view is an interpretive framework organized around persistence, Delta, boundary, accessibility, resolution, consequence, and changed geometry. Its present value is conceptual and heuristic.

Universal Energy Dynamics, Structured Vacuum Energy, Alignment Threshold Time, and deeper resolution geometry are proposed components of a developing foundational programme. They are not established parts of thermodynamics or accepted physical consensus.

Structured Vacuum Energy should not be confused with the quantum vacuum merely because both discussions concern foundational physical reality. A valid connection would require explicit mathematical formulation and comparison with established quantum field theory and observation.

Alignment Threshold Time should not be confused with activation energy, relaxation time, critical temperature, nucleation threshold, response time, or another established parameter. Its independent operational meaning remains to be developed.

Interdisciplinary comparisons in this book identify recurring questions about difference, constraint, pathway, and persistence. They do not establish that cells, engines, atmospheres, crystals, and stars operate through one already demonstrated mechanism.

The Shared-Reality Principle governs the framework. Chavanian interpretation and conventional science must confront the same observations. No empirical quantity may be changed, dismissed, or redefined merely to protect the interpretation.

If future mathematical developments derived from U E D conflict with reliable observation, those developments must be revised or rejected. Conceptual elegance, explanatory comfort, and authorial commitment do not override the empirical world.

The originality claimed by the Chavanian framework lies in its proposed organization and interpretation, not in ownership of the established thermodynamic laws, quantities, experiments, historical discoveries, or scientific traditions discussed throughout the book.

Readers should distinguish four levels at all times: observed event, established scientific description, Chavanian interpretation, and possible future testable claim. The book has succeeded only if those levels are easier to distinguish at the end than they were at the beginning.

Don't miss out!

Visit the website below and you can sign up to receive emails whenever Sandeep Chavan publishes a new book. There's no charge and no obligation.

https://books2read.com/r/B-A-EPGPC-DAIQJ

BOOKS 2 READ

Connecting independent readers to independent writers.

Did you love *What If Change Has an Architecture?*? Then you should read *Delta – The Structure of Change*[1] by Sandeep Chavan!

[2]

What if reality is not driven by force, intention, or hidden agents—but by structure?

In **Delta**, Sandeep J. Chavan introduces a disciplined and accessible framework for understanding change through imbalance, buildup, and resolution. Instead of treating events as isolated causes or outcomes as rewards and punishments, this book reveals a deeper structural logic beneath experience.

---

1. https://books2read.com/u/mKn5nv
2. https://books2read.com/u/mKn5nv

There is no privileged observer in this model.No mystical explanation when outcomes feel uncomfortable.No reliance on abstraction detached from lived reality.

Delta reframes physics, perception, and human systems through a resolution-based lens. Change is not something that "happens" to reality—it is how reality reorganizes itself when structure can no longer sustain imbalance.

Drawing from systems thinking, thermodynamics, structural causality, and Universal Energy Dynamics (UED), the book dismantles force-based interpretations of motion, conflict, and consequence. It replaces them with a cleaner model rooted in equilibrium, dissipation, and structural thresholds.

This is not a speculative cosmology.It is not metaphysical belief.It is not mathematical intimidation.

It is a shift in perception.

Readers interested in physics, philosophy of science, systems theory, consciousness studies, and applied structural thinking will find Delta both challenging and clarifying. Engineers, educators, thinkers, and curious general readers will discover a practical framework that scales from cosmic behavior to everyday human experience.

By the end of this book, you may begin asking different questions:

Not "Who caused this?"But "What built this?"

Not "Why did this happen to me?"But "What structure made this resolution inevitable?"

Delta does not promise control.It offers readability.

And readability changes everything.

Read more at https://www.sandeepjchavan.com/.

www.ingramcontent.com/pod-product-compliance
Lightning Source LLC
LaVergne TN
LVHW100728061126
851857LV00011B/436
* 9 7 9 8 2 3 5 4 8 9 4 9 3 *